PROBLEMS AND QUESTIONS IN PHYSICS

PROBLEMS AND QUESTIONS IN PHYSICS

By

Charles P. Matthews

Joen Shearer

[1897]

Norwood Press
J. S. Cushing & Co. — Berwick & Smith
Norwood Mass. U.S.A.

PREFACE

THERE is perhaps little that need be said prefatory to a work
of this character. The class-room experience of the authors
leads them to believe that any text in Physics needs to be sup-
plemented by problem work in considerable variety. A numerical example in Physics serves a manifold purpose. It
takes the mathematical expression of a physical law out of
the realm of mere abstraction, by emphasizing the connection
between such a law and the phenomena of daily observation.
At the same time, it gives the student an idea of the relative
magnitude of physical quantities and of the units in which they
are measured. Lastly, it shows him the usefulness of his
previously acquired mathematical knowledge, while impressing
upon him the limitations which must be put upon this knowledge when applied to physical relations. There would seem,
therefore, to be no lack of justification for the not inconsiderable
labor of writing an extensive series of problems.

In the preparation of the following pages, the authors have
introduced a number of features which have seemed good to
them, and, it is hoped, may meet with general favor. The
problems are numbered consecutively throughout the book in
Arabic numerals. The paragraphs of the Introduction are numbered in Roman numerals. This contributes to easy reference.
All tables of physical constants are placed in the Introduction.
To work the problems it will be necessary, not only to *read* the
Introduction, but to refer to it continually. The authors con-
fess that in this arrangement they have aimed to abolish the

idea, prevalent in the student mind, that an "Introduction," like a "Preface," is something that no one ever reads. The plan also shortens the statement of a problem, relieving it of much reiterated information.

A few words should be said concerning the use of the calculus notation. As the tendency of writers of elementary works in Physics seems to be towards a greater use of the language of the calculus, it is only appropriate that a fair number of problems should be inserted here which cannot be satisfactorily worked by other than calculus methods. Their number, however, is not large, and the usefulness of the book to students not prepared for them will be in nowise diminished. It is believed that the number of problems is sufficiently large to enable the instructor to make an adequate selection for any class.

Occasional questions not requiring numerical answers have been asked. These are purposely few in number, and are put in to indicate the general character of class-room and examination questions, and with no thought of encroaching upon the province of the instructor.

Here and there graphic methods have been suggested which may be profitably extended by the student. On the other hand, solutions and hints have been omitted in many cases where the student might perhaps expect to find them. It is felt that the methods preferred by the instructor in charge or suggested by the text in use should be used rather than those of the writers, since the general character of the course and the degree of the student's advancement may be thus considered. It is not expected that the student should work the problems without suggestion, and inability to do so in particular cases may indicate to both student and instructor just where some law or definition is not clearly understood. There are undoubtedly obscurities in the text and errors in answers, and the authors would esteem it a favor if readers would call attention to them.

Some criticism may be incurred because of the use of mixed units. Many of the students who will use these problems are pursuing engineering courses. In such case they must of necessity use engineering units. The aim has been not so much to train them in the use of these units, — an abundance of this training comes to them during their course, — but to bring out the relation of the so-called "practical" and gravitational units to the C.G.S. units of Physics.

Suggestions have been received from many sources, among others the works of Jones, Jessop, and Everett. The authors' thanks are due to Messrs C. D. Child, C. E. Timmerman, and O. M. Stewart, Instructors in Physics at Cornell University, for solutions of problems and many valued suggestions.

DECEMBER, 1896.

CONTENTS

	PAGE
MEASUREMENT AND UNITS	1
PHYSICAL TABLES	12
DIRECTED QUANTITIES	21
GRAPHIC METHODS	26
AVERAGES	31
APPROXIMATIONS	33
MECHANICS OF SOLIDS	37
LIQUIDS AND GASES	89
HEAT	100
ELECTRICITY AND MAGNETISM	121
SOUND AND LIGHT	191
MATHEMATICAL TABLES	225
ANSWERS	237
INDEX	245

PROBLEMS IN PHYSICS

I. INTRODUCTION

Measurement. — Whenever, in the domain of physical science, we step from the position of a mere observer of the phenomena around about us to that of an investigator, we seek the aid of a process known as measurement. Whether this process be simple or complex, there is but one operation in it that is fundamental, — the determination of the value of one magnitude in terms of another of the same kind. We may content ourselves with the crudest approximation, — as when we estimate mountain heights in terms of the highest peak of the range, — or, we may make a comparison with the utmost scientific accuracy, using for such a purpose a quantity agreed upon among men as a standard or *unit*. In either case the result sought is a ratio ; namely, that existing between the magnitude and the chosen unit of like kind. This ratio is the *measure* of the given magnitude, and the process by which it is found is called *measurement*. The accuracy with which measurements are made is governed largely by practical needs. It should, however, be borne in mind that the process is, at the best, an approximate one. Even the most exact measurements of physics must be regarded as attempts to determine numerical quantities whose true values must ever remain unknown.

Units. — It follows that the complete expression of a physical quantity, so far as its magnitude is concerned, involves two factors, one a concrete unit, the other a number or *numeric*. Thus if L be a unit of length, the measure or numerical value of a length l is $n = \dfrac{l}{L}$, and the complete expression of the magnitude of l is

$$l = nL.$$

The product of numeric and unit is constant. Whether a debt be paid in dimes or in dollars, it is yet the same debt, but the number representing it in the one case is ten times that representing it in the other. The unit and numeric, in other words, vary inversely.

Fundamental and Derived Units. — Consider the case in which the unit of length is taken as the foot, and the unit of area the square yard. Then a rectangular area a feet long and b feet wide is expressed as

$$A = \tfrac{1}{9} ab \text{ sq. yd.}$$

And, in general, the area is given by

$$A = kab,$$

where k is a constant depending upon the units of length and area involved. If, however, it is agreed that the unit of area shall be the square foot, the value of k reduces to unity, and

$$A = ab \text{ sq. ft.}$$

It thus appears that, in a system made up of arbitrarily chosen units, transformations call into use a number of proportionality constants, many of which will involve endless decimals, introducing into computations much unnecessary labor and liability of error. The earlier units were largely of this character. They were chosen to meet the needs of practical life at a time when simple and definite relations among them were not deemed essential. Thus the origin of the foot is obvious, as is also its variation in different countries.* Further,

---

* The Russian foot is 30.5 cm.; the Austrian foot, 31.6 cm.; the Saxon foot, 28.32 cm.; etc.

derived units based on powers of the fundamental are not always convenient. The yard is a convenient length for the measurement of cloth, but the cubic yard is too large a volume for the grocer's needs. Yet the awkwardness of systems made up of grains, scruples, drams, and ounces, of links, rods, and chains, needs no comment. The metric system, now generally used by physicists, obviates these difficulties by making all change ratios multiples or sub-multiples of 10. All the complex units of physics are thus bound together by ties that may be easily manipulated.

The system in common use is based on three arbitrarily chosen units. These are

the *centimeter*, the $\frac{1}{100}$ part of the length of a certain platinum bar kept in the Archives of Paris ;

the *gram*, the $\frac{1}{1000}$ part of a certain piece of platinum (*the kilogram des Archives*) which is intended to have the same mass as a cubic decimeter of water at the temperature of maximum density (3.9° C.) ;

the *second*, the $\frac{1}{86400}$ part of the mean solar day.

These units of length, mass, and time, respectively, are known as the *fundamental* units of the C.G.S. system. Other units based upon them are called *derived* units.

Another system, much less in use, is based on the same physical quantities, but the units of length and mass are of different value. They are

the *foot*, as a unit of length ;
the *pound*, as a unit of mass ;
the *second*, as above defined.

These units are the basis of the foot-pound-second (F.P.S.) system of units.

Referring again to the equation

$$A = kab,$$

we see that in the C.G.S. system in order to make k unity the unit of area must be taken as the square centimeter. The resulting equation is

$$A = ab,$$

concerning the reading of which a word of caution is necessary. When fully translated it affirms that the *number* of units of area is equal to the *number* of units of length × the *number* of units of breadth. In other words, it is the numerics that are actually multiplied. So, force is measured by the acceleration produced in mass. The equation

$$F = ma$$

is usually read *force equals mass times acceleration.* This is an abbreviated statement of the fact that, in a consistent system of units, the number of units of force equals the number of units of mass × the number of units of acceleration produced.

Velocity is the rate of motion. The units of length and time being the *centimeter* and the *second*, any other unit of velocity than the *centimeter per second* is both awkward and unscientific. Similarly the C.G.S. unit of acceleration must be an acceleration such that unit velocity is gained in one second. Acceleration is measured, therefore, in centimeters per second per second.

The more complex electrical and magnetic units are built up in the same natural way. It is found that the force between two magnetic poles varies as the product of their pole strengths and inversely as the square of the distance between them. That is, in air,

$$F = \frac{mm'}{r^2}.$$

Whence *unit magnet pole* is a pole of such strength that it repels an equal and like pole, placed 1 cm. away, with a force of one dyne. This unit of force, itself a derived unit, has already been referred to.

Dimensions and Dimensional Equations. — Suppose that for the unit of area in any system a square be taken one of whose sides is the unit of length, and let an area a contain n such units. That is,

$$a = nA.$$

Further suppose that it is desired to double the unit of length. The new unit of area built upon the changed unit of length is four times the old unit. In other words, the unit of area varies as the square of the unit of length, or it is said to be of two *dimensions* in length. To indicate this, the last equation may be written

$$a = nL^2.$$

Let v be a concrete velocity such that a distance l is traversed in time t. The numerical values of these quantities are found by dividing each by the appropriate unit. Let V, L, and T be these units. Then the numerical values are $\dfrac{v}{V}$, $\dfrac{l}{L}$, $\dfrac{t}{T}$. We have then two numerical values of this velocity, viz.,

$$\frac{v}{V} \quad \text{and} \quad \frac{\dfrac{l}{L}}{\dfrac{t}{T}},$$

but these values are to be equal, which gives

$$\frac{v}{V} = \frac{l}{L} \cdot \frac{T}{t}.$$

Writing the equation so as to separate the units' part, we have,

$$V = \left(v\frac{t}{l} \right) \frac{L}{T}.$$

Or, in words, the unit of velocity varies directly as the unit of length and inversely as the unit of time. That is, the dimensions of unit velocity are LT^{-1}. In passing to dimensional equations we may discard constant numerical factors, since the units, and therefore the dimensions, are not affected thereby.

So, the dimensions of the unit of acceleration are readily seen to be LT^{-2}; of the unit of force, MLT^{-2}; of the unit of work, ML^2T^{-2}; and so on.

It becomes apparent at once that dimensional formulas show the powers of the fundamental units that enter into derived units. Hence dimensional equations are of much use in facilitating change of units.

EXAMPLE. The numerical value of the acceleration due to gravity, when the centimetre and second are used as units of length and time, is 980. Find the value in terms of the foot and minute.

The dimensions of acceleration, it has been seen, are LT^{-2}. We have

$$g = 980 \left[\frac{\text{cm.}}{\text{sec.}^2} \right]$$

$$= 980 \left[\frac{.033 \text{ ft.}}{(\frac{1}{60} \text{ min.})^2} \right]$$

$$= 980 \times .033 \times 3600 \left[\frac{\text{ft.}}{\text{min.}^2} \right]$$

$$= 116424 \left[\frac{\text{ft.}}{\text{min.}^2} \right].$$

That is to say, the acceleration due to gravity is 116424 ft. per minute per minute.

Whenever problems involving change of units occur in the following collection, the student is strongly advised to work them in this way, until the processes become so familiar as not to need formal statement.

The two members of every equation must reduce to the same dimensions, otherwise the equation is absurd. Or, what amounts to the same thing, every term of an equation is homogeneous with respect to each fundamental unit involved. The equation of the motion of a particle having uniform acceleration in the direction of motion is

$$l = a + bt + ct^2,$$

wherein l and a are of the dimensions L,

· b, a velocity, is of the dimensions LT^{-1},

and c, an acceleration, is of the dimensions LT^{-2}.

Thus each term of the expression for l is of the dimensions L of l itself.

This gives a very convenient check upon our work in deriving such an equation.

Mass and Weight. — These words stand for two distinct physical concepts. Thus, mass is *quantity of matter*, while weight is *force*. Physically, then, they are no more alike than length and time. Not infrequently the beginner fails to apprehend this fact. Confusion arises partly because masses are compared by comparing their weights, and partly because the same word is often used ambiguously to name both a unit of mass and a unit of force.

If a point move over equal spaces in equal times, any constant distance corresponds to a constant time. Or, in other words, distance traversed and time vary in direct proportion. For example, when, in railroad parlance, two stations are said to be "four hours" apart, every one understands roughly what *distance* is meant. Now it is precisely this relation that exists between mass and weight, and it is largely because of their proportionality in any one locality that some license is admissible in naming their units.

Masses attract each other according to the fundamental law of gravitation. To the attraction between the earth and the bodies upon its surface the special name *weight* is given. The weight of a body, therefore, is the force with which it is drawn towards the earth, or with which the earth is drawn towards it. When two bodies are placed in opposite pans of a beam balance and do not destroy its equilibrium, they are said to be of equal weight. That is, the forces acting at the ends of the beam are equal. Further, by the law of proportionality, the bodies are of equal mass, since we have for each force (or so much of it as may be due to the added mass),

$$F = Mg,$$

wherein g is the acceleration with which the mass M would fall if released. The balance thus serves to determine equal masses, and it is evident that if the system were carried to any other locality the equilibrium would remain perfect, the masses remaining unaltered and the weights varying with g. It is in this way that masses are compared through the agency of their weights.

As to units of mass, there are two in common use:

the *pound*,
the *gram*,*

each of which is the quantity of matter in a certain carefully preserved piece of platinum. To obtain the weights of these masses we must multiply by the value of g appropriate to the system of which the unit is a fundamental, and to the locality at which the weight is desired. Thus the weight of a pound where $g = 32.2$, is

$$W_p = mg = 1 \times 32.2$$
$$= 32.2 \text{ units of force in the F.P.S. system}$$
$$= 32.2 \text{ } poundals.$$

The weight of a gram where $g = 980$ is

$$W_g = mg = 1 \times 980$$
$$= 980 \text{ units of force in the C.G.S. system}$$
$$= 980 \text{ } dynes.$$

All this is clear enough. But unfortunately, perhaps, the terms *pound* and *kilogram* are used in such expressions as, "a body weighs 16 pounds" or "a weight of 12 kilograms." The

---

* The original standard is the *kilogram* = 1000 grams.

pound and kilogram being units of mass, such usage, taken literally, is absurd. The expressions, however, are elliptical, their full meaning being "a body weighs the same as 16 pounds weigh," or "a weight equal to the local weight of 12 kilograms." Or, we may say, with equal correctness and greater brevity, "16 pounds' weight" or "12 kilograms' weight." So, a grocer is said to weigh out tea; but he does not sell *weight* — he has no *force* for sale — but *mass*.

A still greater source of confusion arises from the fact that the engineer finds the poundal ($\frac{1}{32.2}$ pound's weight)* and the dyne ($\frac{1}{980} \times \frac{1}{1000}$ kilogram's weight) too small for practical needs as units of force. The engineering unit of force among English-speaking people is the weight of a pound (called simply *a pound*), and among people using the metric system the weight of a kilogram (called simply *a kilogram*). Since these units depend on the value of g, they are slightly variable, but the variation is so small as to be usually negligible for engineering purposes.

As illustrating this last usage, suppose that the piece of platinum which the English people have agreed to call a pound were hitched to a spring balance and the whole arrangement carried to different points on the surface of the earth. The registry of the balance would evidently vary to a slight extent. The engineer says we will neglect this variation as being of negligible importance, and say that any agent which stretches the balance spring ten times as much as does the freely suspended pound mass is a *force* of 10 lb. Let us suppose, then, that in this way a body is found to weigh 10 lb., and let us inquire what the mass of this body is. By Newton's second law this force is measured by the mass of the body times the acceleration which it would possess if allowed to fall freely. Taking $g = 32.2$, we write

* The accepted value of g at Ithaca is 980, which corresponds to 32.15 in foot-second units. 32.2 is commonly used, however. See Church's "Mechanics of Engineering."

$$G = mg,$$

$$10 = m \times 32.2,$$

whence, $$\text{mass} = \frac{G}{g} = \frac{10}{32.2}.$$

This makes the mass of the body invariable, as it must be. To the unit mass in this system no name has been given, but it is readily seen to be the mass of a body weighing 32.2, or more generally g, pounds. With this understanding it is quite correct to say that a body weighs G pounds, to speak of a pull or thrust of G pounds, a pressure of G pounds per square inch, etc.

The pound and the kilogram are sometimes called *gravitational units* of force. Likewise the foot-pound and the kilogram-meter are gravitational units of work, and the horse-power is a gravitational unit of power.

As illustrating this system we may consider the following problems :

A body weighing 12 lb. is moving with a velocity of 193.2 ft. per second. What constant force must be applied to bring it to rest in 3 sec.?

The acceleration is

$$\frac{193.2}{3} = 64.4 \text{ ft. per second per second.}$$

The mass of the body must be found. Since the weight 12 lb. would produce an acceleration of 32.2 ft. per second per second, if the body were allowed to fall, we have

$$12 = m\, 32.2,$$

$$m = \frac{12}{32.2}.$$

Finally $$F = ma$$

$$= \frac{12}{32.2} \times 64.4 = 24 \text{ lb.}$$

A force of 12 kg. is overcome through a distance of 20 m. Find the work done.   We have

$$W = Fl$$

$$= 12 \times 20$$

$$= 240 \text{ kilogram-meters.}$$

This result is dependent on the value of g at the place at which the work is done.

The physicist solves this problem as follows :

A force equal to 12 kg. weight where $g = 980$ is

$$F = 12 \times 10^3 \times 980 \text{ dynes,}$$

and the work done is

$$W = Fl = 12 \times 2 \times 980 \times 10^6 \text{ ergs.}$$

a. What two elements are necessary for the complete expression of the magnitude of a physical quantity? Explain fully in what the process of measurement consists.

b. What is the logical objection to a system of units in which the inch is taken as the unit of length, the square rod as the unit of area and the cubic metre as the unit of volume?

c. A certain surface is a units long and b units wide; the general expression for the area is

$$A = kab.$$

Under what conditions will the area be expressed as ab simply?

d. If in the last example a and b are given in feet, what will be the value of k if the unit of area be taken as 1 square mile?

e. Explain what is meant by fundamental and derived units.

f. Imagine the unit in which a definite magnitude is measured to vary continuously.   Plot values of the unit as abscissæ and corresponding values of the numeric (or measure) as ordinates.   Discuss the locus.

NOTE. — Many examples involving change of units, use of dimensional equations and like matters are to be found further on in this book. It has seemed better to place such examples, with the exception of the few general ones above, where they may be used after the student is in some degree familiar with the ideas involved.

UNITS OF LENGTH.

NOTE. — The student is advised to study the approximate values. They are of assistance in mental calculations, and are frequently sufficiently exact for problem work.

		Roughly approximate values.
1 in.	= 2.54 cm.	$2\frac{1}{2}$.
1 ft.	= 30.48 cm.	$30\frac{1}{2}$.
1 mi.	= 160933 cm.	
	= 1.6 km.	
1 cm.	= .394 in.	$\frac{2}{5}$.
1 cm.	= .0328 ft.	$\frac{3}{100}$.
1 m.	= 39.37 in.	40.
1 km.	= .6214 mi.	$\frac{5}{8}$.

UNITS OF AREA.

1 sq. in.	= 6.45 sq. cm.	$6\frac{1}{2}$.
1 sq. ft.	= 929.01 sq. cm.	
1 sq. mi.	= 25899 × 10² sq. m.	
1 sq. cm.	= .155 sq. in.	$\frac{2}{13}$.
	= .001076 sq. ft.	
1 sq. m.	= 3.861 × 10⁻⁷ sq. mi.	

UNITS OF VOLUME.

1 cu. in.	= 16.387 cu. cm.	$16\frac{1}{2}$.
1 cu. ft.	= 28316. cu. cm.	
1 gal.	= 4541. cu. cm.	
	= 4.54 litres	$4\frac{1}{2}$.
1 cu. cm.	= .061 cu. in.	$\frac{1}{16}$.
	= 3.532 × 10⁻⁵ cu. ft.	

Units of Mass.

1 lb.   = 453.59 g.
1 oz. (av.) = 28.35 g.
1 g.   = 15.43 gr.
            = .0353 oz.
            = .0022 lb.

Units of Force.

[$g = 980$ in all gravitational units.]
1 poundal   = 13825 dynes.
1 gram's weight   = 980 dynes.
1 pound's weight   = 444518 dynes.
1 kilogram's weight = 9.8×10^5 dynes.
1 dyne   = 2249×10^{-9} pound's weight.

Units of Work.

1 foot-pound   = 1.35485×10^7 ergs
                     = 13825 gram-centimeters
                     = .138 kilogram-meters.
1 kilogram-meter   = 7.233 foot-pounds.
1 joule   = 10^7 ergs.
1 watt-hour   = 36×10^9 ergs.
1 horse-power-hour   = 26856×10^2 joules.

Units of Power.

1 horse-power   = 746 watts
                     = 746×10^7 ergs per second
                     = 33000 foot-pounds per minute.
1 watt   = 10^7 ergs per second.

Units of Stress.

1 lb. per square foot = .48826 grams per square centimeter
                      = 478.5 dynes per square centimeter.
1 lb. per square inch = 70.31 grams per square centimeter
                      = 68904 dynes per square centimeter.
1 in., mercury at 0° = 34.534 grams per square centimeter.
1 cm., mercury at 0° = 13.596 grams per square centimeter.

The Mechanical Equivalent of Heat.

$$1 \text{ g. through } 1° \text{ C. } = 4.2 \times 10^7 \text{ ergs}$$
$$= .4281 \text{ kilogram-meters.}$$
$$1 \text{ lb. through } 1° \text{ F. } = 1.058 \times 10^{10} \text{ ergs}$$
$$= 780.8 \text{ foot-pounds.}$$

Table I

DENSITIES

Solids

Aluminum	2.6	Asbestos	2.4
Antimony	6.7	Chalk	2.3–3.2
Bismuth	9.8	Coal	1.4–1.8
Brass	8.4	Cork	.14–.3
Copper	8.9	Glass, common	2.5–2.7
Gold	19.3	Glass, flint	3–3.5
Iron	7.8	Ice	.917
Lead	11.3	Iceland Spar	2.75
Nickel	8.9	Ivory	1.9
Platinum	21.5	Marble	2.7
Silver	10.5	Paraffine	.87–.91
Sodium	.98	Quartz	2.65
Tin	7.3	Oak	.7–1
Zinc	7.1	Pine	.5

Liquids, 0° C.

Alcohol	.806	Sea Water	1.026
Ether	.736	Sulphuric Acid	1.85
Carbon Bisulphide	1.29	Nitric Acid	1.56
Glycerine	1.27	Hydrochloric Acid	1.27
Mercury	13.596	Oil of Turpentine	.87

Table II

Specific Heats of Solids

Aluminum	.2122	Calcium	.1804
Bismuth	.0298	Carbon, diamond	.1128
Brass	.0940	Carbon, graphite	.1604

Carbon, charcoal	.1935	Magnesium	.2450
Copper	.0933	Nickel	.1092
Gold	.0316	Platinum	.0323
Glass	.1877	Silver	.0559
Ice	.5040	Tin	.0559
Iron	.1124	Zinc	.0935
Lead	.0315		

TABLE III

SPECIFIC HEATS OF LIQUIDS

Alcohol	·55
Carbon Bisulphide	.24
Ether	·53

SPECIFIC HEATS OF GASES AND VAPORS

(Constant Pressure)

Air	.237
Oxygen	.217
Hydrogen	3.4
Nitrogen	.244
Steam	.48
Marsh Gas	·593
Alcohol	·453

TABLE IV

MELTING-POINTS AND HEATS OF LIQUEFACTION

	Melting-point. °	Heat of Liquefaction. Calories.		Melting-point. °	Heat of Liquefaction. Calories.
Aluminum	600		Mercury	−40	2.82
Copper	1054		Nickel	1450	4.64
Glass	1100		Platinum	1775	27.2
Gold	1045		Silver	954	24.7
Ice	0	80	Tin	230	14.6
Iron	1600		Zinc	412	28.1
Lead	326	5.4			

TABLE V

BOILING-POINTS AND HEATS OF VAPORIZATION

	Boiling-point. °	Heat of Vaporization. Calories.
Alcohol	77.9	202.4
Bromine	58	45 6
Ether	34.9	90.4
Mercury	350	62
Water	100	536

TABLE VI

UNITS OF HEAT

	Ergs.
1 calorie (gram-degree C.)	$= 4.2 \times 10^7$
1 major calorie (kilogram-degree C.)	$= 4200 \times 10^7$
1 pound-degree Centigrade	$= 1905 \times 10^7$
1 pound-degree Fahrenheit	$= 1058 \times 10^7$

TABLE VII

COEFFICIENTS OF LINEAR EXPANSION

Brass	180	
Copper	170	
Glass	085	
Gold	150	
Iron	120	
Lead	280	$\times 10$
Platinum	085	
Silver	190	
Tin	200	
Zinc	290	

COEFFICIENTS OF VOLUME EXPANSION

Alcohol (mean 0° − 78°)	.00104
Mercury (mean 0° − 100° C.)	.000182
Water (mean 0° − 100°)	.000062

TABLE VIII

THERMAL CONDUCTIVITIES

	Relative Conductivity.	C.G.S.
Silver	100	1.3
Copper	74	0.99
Iron	12	0.16
Lead	8.5	0.11
Bismuth	1.8	0.02
Ice	0.2	0.003
White Marble	0.1	0.001
Glass	0.05	0.0007

TABLE IX

COLLECTED DATA FOR DRY AIR *

Expansion from 0° to 100° at constant pressure as	273 : 373
Specific Heat at constant pressure	.2375
Specific Heat at constant volume	.1691
Standard barometric height	76 cm.
Density at 0° and 76 cm.	.001293 g.
Volume 1 g. at 0° and 76 cm.	773.3 c.c.

* Everett.

TABLE X

RESISTANCE

Substance.	Specific Resistance.	Temperature Coefficient (0–100°).
Aluminum (annealed)	$289 \cdot 10^{-8}$ ohms	. . .
Copper (annealed)	$160 \cdot 10^{-8}$ ohms	$388 \cdot 10^{-5}$
Gold	$208 \cdot 10^{-8}$ ohms	$365 \cdot 10^{-5}$
Iron (pure)	$964 \cdot 10^{-8}$ ohms	. . .
Iron (telegraph wire)	$1500 \cdot 10^{-8}$ ohms	. . .
Lead	$1963 \cdot 10^{-8}$ ohms	$387 \cdot 10^{-5}$
Mercury	$9434 \cdot 10^{-8}$ ohms	$72 \cdot 10^{-5}$
Platinum	$898 \cdot 10^{-8}$ ohms	. . .
Silver	$149 \cdot 10^{-8}$ ohms	377
German Silver	$2100 \cdot 10^{-8}$ ohms	$44 \text{ to } 65 \cdot 10^{-5}$
Platinoid	$3200 \cdot 10^{-8}$ ohms	$21 \cdot 10^{-5}$
Manganin	$4700 \cdot 10^{-8}$ ohms	$122 \cdot 10^{-5}$

c

TABLE XI

UNITS OF RESISTANCE

1 true ohm unit of resistance.　　　1 B. A. unit　　= .9867 true ohms.
1 legal ohm = .9972 true ohms.　　1 Siemen's unit = .9407 true ohms.

TABLE XII

SPECIFIC INDUCTIVE CAPACITIES

Air = 1

Solids.	K.	Liquids.	K.
Glass	4 to 7	Acetone	21.8
Gypsum	5.6	Alcohol	25
Ice	2.85	Aldehyde	18.6
Iceland Spar	7.4	Benzine	2.3
Marble	6.4	Carbon Disulphide .	2.5
Mica	6 to 8	Ether	4.27
Paraffine	2.2	Glycerine	56.2
Quartz	4.54	Oils	2.2
Rosin	2.55	Petroleum	2.06
Rubber, soft	2.4	Turpentine	2.23
vulcanite . .	2.7	Water	75.5
Salt	5.8		
Sandstone	6.2	Gases.	
Shellac	3	Hydrogen	0.9998
Sulphur	2.69	Vacuum	0.9985
Wood	2.95	Vapors	1.001 to 1.01

TABLE XIII

PRACTICAL UNITS EXPRESSED IN C.G.S UNITS

Let V be the velocity of light, about $3 \cdot 10^{10}$ cm. per sec.

	Electromagnetic		Electrostatic. C.G.S.	
	Practical.	C.G.S.		
Quantity	1 coulomb	$1/10$	$V/10, i.e.$	$3 \cdot 10^9$
Current	1 ampere	$1/10$	$V/10$	$3 \cdot 10^9$
Potential	1 volt	10^8	$10^8/V$	$1/(3 \cdot 10^2)$
Resistance	1 ohm	10^9	$10^9/V^2$	$1/(9 \cdot 10^{11})$
Capacity	1 farad	$1/10^9$	$V^2/10^9$	$9 \cdot 10^{11}$
Self-induction . . .	1 henry	10^9		

TABLE XIV

SOUND

VELOCITY OF SOUND IN METERS PER SECOND

Solids (20° C.).		Liquids (20° C.).		Gases (o).	
Brass	3480	Alcohol . . .	1160	Air	332
Copper . . .	3560	Water . . .	1440	Illuminating Gas,	490
Iron	5130	Petroleum . .	1395	Hydrogen . . .	1280
Steel, cast . .	5000			Oxygen . . .	317

TABLE XV

LIGHT

Velocity of light $\begin{cases} 299860 \text{ kilometers per sec.} & \text{Nearly } 3 \cdot 10^5. \\ 186323 \text{ miles per sec.} \end{cases}$

TABLE XVI

WAVE-LENGTHS OF THE PRINCIPAL FRAUNHOFER LINES IN
TENTH-METERS *

Line.	Wave-length.	Line.	Wave-length.
A	7594.059	M	$\begin{cases} 3727.763 \\ 3727.20 \end{cases}$
B	6867.461		
C	6563.054	N	3581.344
D_1	5896.154	O	3441.135
D_2	5890.182	P	3361.30
E	$\begin{cases} 5270.533 \\ 5270.448 \\ 5269.722 \end{cases}$	Q	3286.87
		R	$\begin{cases} 3181.40 \\ 3179.45 \end{cases}$
F	4861.496	S_1	3100.779
G	$\begin{cases} 4308.071 \\ 4307.904 \end{cases}$	S_2	3100.064
		T	$\begin{cases} 3021.191 \\ 3020.759 \end{cases}$
H	3968.620		
K	3933.809	U	2947.993
L	3820.567		

* 1 tenth-meter = 10^{-8} of a centimeter.

TABLE XVII

INDICES OF REFRACTION [D LINE] *

	Density.	Index.
Glass (hard crown)	2.486	1.517
Glass (soft crown)	2.55	1.5145
Glass (light flint)	3.206	1.574
Glass (dense flint)	3.658	1.622
Glass (extra dense flint)	3.889	1.65
Glass (double extra dense flint)	4.429	1.71
Rock Salt	. . .	1.544

* Everett, C.G.S. Units and Constants.

LIQUIDS

Alcohol 1.363	Ether 1.36
Canada Balsam 1.54	Olive Oil 1.47
Carbon Disulphide . . . 1.63	Turpentine 1.48
Chloroform 1.446	Water 1.334

UNIAXIAL CRYSTALS

	Ordinary Index.	Extraordinary Index.
Iceland Spar	1.6584	1.4864
Tourmaline	1.6366	1.6193
Quartz	1.5432	1.5512

TABLE XVIII

NUMERICAL CONSTANTS

LOGARITHMS

$\epsilon = 2.7183$ . . . . $\text{Log}_{10} \epsilon = .434294$

$\text{Log}_{10} N = \text{Log}_\epsilon N \cdot .434294$

$\text{Log}_\epsilon N = \text{Log}_{10} N \cdot 2.302585$

1 radian . . . . $57°.2958$

$1°$ . . . . .01745 radians

		Log$_{10}$			Log$_{10}$
$\pi = 3.14159$		.497149	π^2	9.8696	.994299
π approx. 22:7			$1:\pi^2$ . . .	.10132	1.005700
$1:\pi$ . . .	.3183	1.502850	2π	6.283	.798179
$\sqrt{\pi}$ . . .	1.772	.248575	$1:2\pi$ . . .	1592	1.201820
$1:\sqrt{\pi}$ . . .	.5642	1.751425			
$\sqrt{2}$ . . .	1.4142	.150515	$\sqrt{3}$ . . .	1.7321	.238561
$1:\sqrt{2}$ . . .	.7071	1.849485	$1:\sqrt{3}$ . . .	.5773	1.761439

II. DIRECTED QUANTITIES, VECTORS

Many of the quantities considered in physics involve the idea
of direction, and require the statement of two things before we
can form any clear idea of them. First, we must state how
large they are as compared with a thing of like kind taken as
a unit; second, in what direction they must be taken. The
familiar idea of motion from one point to another may be considered as typical of this class of quantities. Suppose one asks
the way from one point in a city to another. The answer might
be to go a certain distance north, then a certain distance west,
etc. Or, if circumstances permit, he may be told to go a certain
distance in a specified direction without turns.

The answer is one based on the experience that we may go
from one point to another either by a series of connected "steps"
or courses such that they begin at the starting-point and end at
the final one, or by a single step, the straight line joining the
points. Or, since the result is the same so far as change of
position is concerned whether we take the crooked path or the
straight, we may call the latter the **resultant** of the former.

In considering the geometry of the problem, it may be noted
that if we are given the steps 1 and 2 we may (Fig. 1) find
their resultant in either of
two ways: from A we may
lay off 1 in its proper direction, and from the end of
1 lay off 2 in like manner.
The line joining the ends

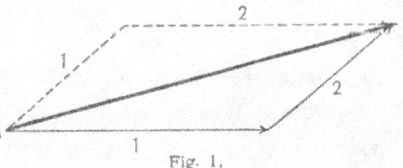

Fig. 1.

of 1 and 2 is then the resultant required. Or we may form
a parallelogram with one corner at A, and whose sides are 1

and 2. The diagonal drawn from A is the equivalent step or resultant.

The student should remember that the problem of finding the resultant of a given system of steps is perfectly definite, and only one solution can be found; but the converse is not true, as a given step may be made up of any one of an endless number of step systems.

The process of finding the resultant of a given system is often spoken of as the **composition** of steps; while that of replacing a single step by a system, usually two, is called the **resolution** of steps.

The simplest, and, at the same time, the most useful case of resolution is when the step is resolved into two at right angles

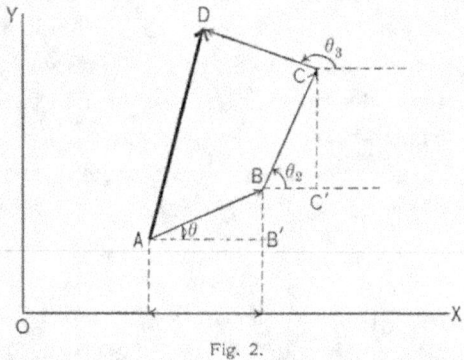

Fig. 2.

to each other. Or the line is said to be **projected** on two rectangular axes X and Y.

Then X component $= AB' = AB \cos \theta$,
            Y component $= BB' = AB \sin \theta$.

The name *vector* (*i.e.* carrier) is usually applied to this class of quantities, and the resultant of a system of vectors is spoken of as the **vector sum** of the components. A thorough understanding of the geometrical ideas involved in adding and resolving vectors is of the greatest importance to the student in physics, and must be acquired before any real progress in the subject is made. The following simple problems are added to assist the student toward this end.

1. Which of the following quantities are vectors? Force; mass; acceleration; momentum; energy; volume; velocity; current; weight; time; interest.

2. Show by diagram the vector sum (*i.e.* the equivalent straight path) of the following set of paths: E. 4 mi.; N. 2 mi.; N.W. 3 mi.; S.W. 5 mi.

3. Draw the same set of paths in the reverse order; *i.e.* S.W. 5 mi.; N.W. 3 mi.; etc.

4. When the vectors are not in the same plane, show how the vector sum is found.

5. What is the vector sum of the length, breadth, and height of a room?

6. Two vectors at right angles to each other, of lengths 4 and 3 respectively, have what vector sum or resultant? If at 60°? 180°? 0°?

7. Six vectors equal in length are placed end to end so that the angle between each pair is 120°. What is the vector sum?

8. Show that the order in which "steps" are taken in no way modifies the sum.

A vector may be given in either of two ways, — by its components or by its length and direction, or the angle it makes with a given line. In the following examples the line of reference is the horizontal line drawn to the right (*x*-axis).

9. Find the resultant of the following vectors: 3, 25°; 4, 100°; 2, 200°; 5, 300°. The work may be conveniently arranged as follows:

Length	Dir.	cos	sin	X comp. $l \cos$	Y comp. $l \sin$
3 . . .	25°				
4 . . .	100°				
2 . . .	200°				
5 . . .	300°				

10. Draw the following vectors : 3, 90° ; 4, 180° ; 5, 190°.

11. Draw the vectors whose components at right angles to each other are 2 and 3, 4 and 6, 2 and − 3.

12. A vector 10 units in length makes an angle of 30° with one of two perpendicular lines. Find the component along each line.

13. A given vector is to be resolved into two at right angles, such that one component is double the other. Find the angle which the longer must make with the given vector.

14. Could the vector AB be considered as the vector sum of the set of short vectors parallel to the axes? Those parallel to X are called what in calculus? Those parallel to Y?

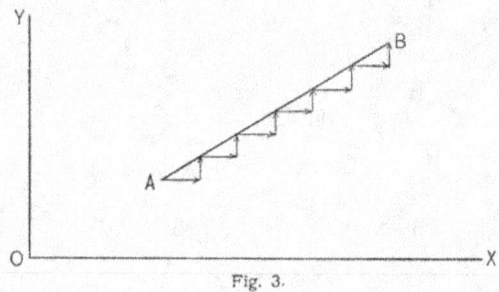

Fig. 3.

15. Two vectors, a and b, are given, making angles θ_1 and θ_2 with the reference line. Find the sum of their X components. Find the sum of their Y components. From these find the

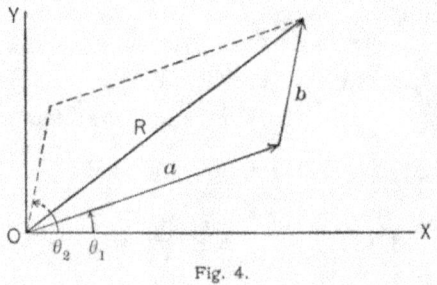

Fig. 4.

resultant of a and b. Reduce to the formula given in trigonometry for the cosine of an angle in terms of the sides.

16. Show that the resultant of two vectors may be found from the theorem in geometry: The square on any side of a triangle is equal to the sum of the squares on the other two sides ± twice the product, etc.

17. n coplanar vectors are drawn from a common point. A polygon is formed by joining their extremities. Prove that the resultant is given in magnitude and direction by n times the vector joining the origin and the center * of gravity of the polygon.

18. Test the above statement for two, three, and four vectors.

19. If the vectors were so numerous that their ends formed a *continuous curve*, what method could be used to find the resultant?

* See 197.

III. GRAPHIC METHODS

It is frequently impossible to keep in mind the complete time history of variable phenomena, or to readily compare the values of quantities which alter with time or position. A clearer conception in such cases may often be obtained by some geometrical method of representing the relative values of two quantities at different times or places. Take, for example, the motion of a ball struck by a bat ; we may wish to compare any two of the various quantities which are involved in its motion. The height above the earth may be compared with the horizontal distance from the starting-point, or with the time since it was struck, or with its vertical velocity, etc.

In the first case, we might draw an actual picture of its path to reduced scale, as (Fig. 5). If we wished to compare height

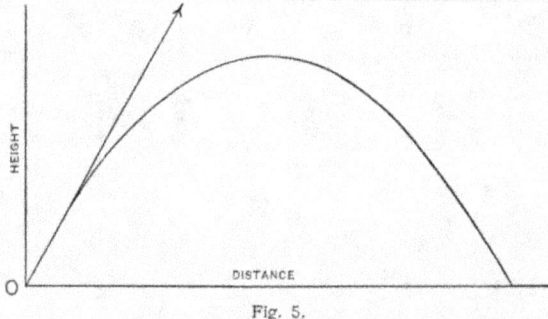

Fig. 5.

at any instant and time since the ball was struck, we might measure a series of lengths to *suitable* scale, along a straight line, to represent heights, and label each with the time required to reach that height. This would, however, be confusing, since the ball is at the same height, in general, twice. Suppose we displace each height h as many arbitrary units to the right as units of time have elapsed since starting, as t_0, t_1, t_2, etc., and the corresponding heights h_1, h_2, h_3, etc. (Fig. 6). We know, however, that the ball took, in succession, every height between those indicated; hence if we were to erect a perpendicular at every point between t_1 and t_2, and measure along each the corresponding height of the ball, the ends of these perpendiculars would form a continuous curve. This process is known as "plotting" the curve, and is of fundamental importance in

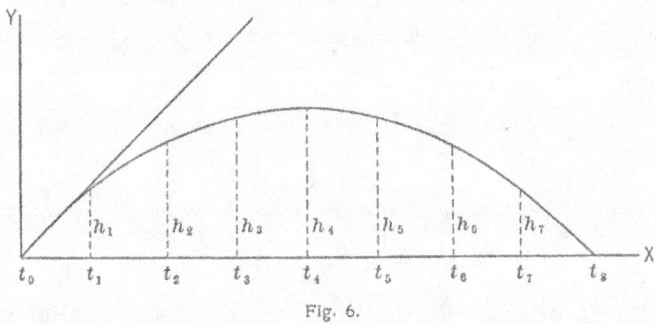

Fig. 6.

the study of physics. The two lines of reference from which distances are measured are called the **axes** of co-ordinates, and are usually chosen at right angles to each other. One is often called the axis of x, and the other the axis of y, and the lengths measured along the x-axis are called **abscissas** or x's. Those measured along or parallel to y are called **ordinates** or y's, and any x with its corresponding y are called the **co-ordinates** of the point which they determine.

"Self-registering" instruments usually draw a curve by some mechanical device. An example is the self-registering thermometer, where a pen is made to rise and fall with the temperature, while the paper is drawn at a uniform rate in a line perpendicular to the motion of the pen. A curve such as the following is the result (Fig. 7). Both time and temperature are continuous, and the curve is a fairly true picture of the time-temperature relation.

In case we had *observed* the temperature at 2, 2.30, 3, 3.30, 4, etc., and had no knowledge of intermediate temperatures, we would *draw* a continuous curve through the *observed* points, which would be more and more reliable as the time intervals were made *smaller*. In general, the more irregular the changes in the observed quantity, the shorter these intervals must be made to ensure that no sudden variation escapes notice.

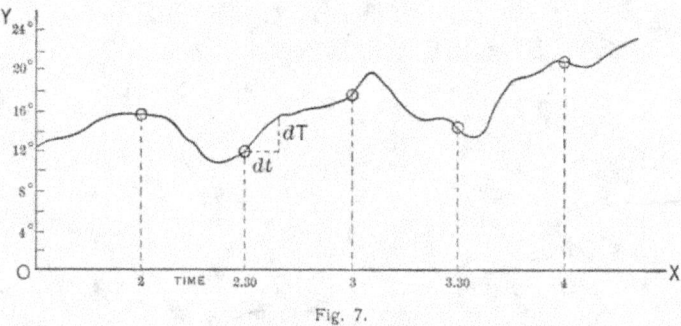

Fig. 7.

We may expect in each case certain peculiarities in the curve, depending on the physical relations which determine it, and, conversely, any peculiarity, as a maximum or minimum, change of curvature, asymptote, etc., will usually have a physical meaning. For example, every change of temperature requires a certain time interval, so that no portion of the time-temperature curve can be vertical. Time never decreases, and temperature has only *one* value at a given instant, so there are no "loops" or multiple points in such a curve.

When we consider the quantity of heat supplied to a gram of ice, for example, and the resulting change of temperature, we find a curve with certain abrupt changes (see Fig. 8). Starting at 0°, 80 heat units are used with no increase of *t*. The line *AB* shows the quantity-temperature relation after melting (approximately straight). At 100° we have an abrupt rise to *C* then, another straight line whose slope is dependent on conditions of pressure, etc. The amount of heat required per gram for any temperature change may be read from the curve.

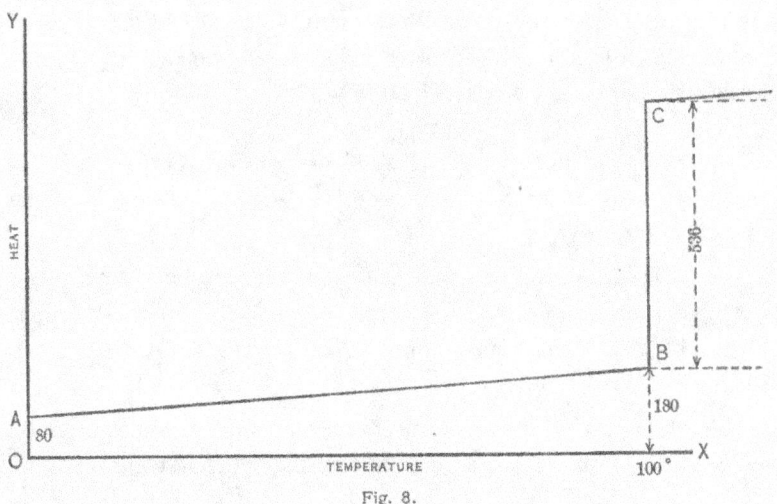

Fig. 8.

Curves are used in physics for various purposes ; as,

(*a*) To represent graphically general laws.

Ex. Path of a projectile. Laws of falling bodies.

(*b*) As a record of results of observation of two related varying quantities.

(*c*) For use in computation. As a sort of numerical map of simultaneous values.

The student should not rest content with simply drawing the curve, but should endeavor to associate the changes or peculiarities in form with the underlying physical conditions. If familiar with the methods of analytic geometry and calculus, he may apply these methods to their study.

In particular, if the curve is a graphic representation of a general law, he should note whether all portions of the curve have an actual physical interpretation, — whether the physical conditions indicated by certain portions of the curve could be realized ; if it cuts the axes, what the intercepts mean ; whether the direction of the tangent line at any point has a physical interpretation ; does the area of a given portion represent some physical quantity ; etc.   When it is drawn from observed values, the relation between the co-ordinates may often be expressed as an algebraic equation, either from its general appearance or from a knowledge of the physical law involved.

20. Draw a curve showing the relation between the side of a square and its area.   Interpret its "slope."   Should it pass through the origin ?

21. Draw a curve showing the relation between simple interest, principal, and time.   What is the slope?   How would the curve of *amount* and time differ from this ?   Interpret the intercepts in this case.

22. Given the curve of displacement and time, how could you find the velocity-time curve ? the acceleration-time curve ?

IV. AVERAGES

When we have to deal with a series of values of the same quantity at different times or places, it is convenient to substitute for the series a single quantity, so chosen that the result will not be changed. Such a quantity is known as an "average" or a mean value. For example, we may wish to consider the temperature of the air at a certain point during a certain period of time, as an hour. Some of this time the temperature may have been rising and some of the time falling, and these changes may have been more or less rapid and irregular. To find the temperature which may fairly be taken to represent the temperature at that point during the hour, we would be obliged to add together a great *number* of observed temperatures and divide the result by this number. The greater the number added, the more nearly correct the average. We might also have required the average temperature at a given instant along a given line, over a given area or throughout a given volume. In all these cases we should take the sum of an indefinitely great number of separate values and divide by the time, length, area, or volume considered. We actually only approximate this by taking a smaller number. The actual addition of these quantities can in certain cases be avoided. As when the values to be averaged increase or decrease at a constant rate, the terms then form an arithmetic series, and the mean value is one-half the sum of the first and last. Examples of this will be found in problems on velocity, force, etc. Again, when a curve is drawn showing the relation between the two variables, if by means of calculus or otherwise we are able to find the area $ABB'A'$, we may divide this area by AB and get the *average ordinate.*

For,   Area $= \int y dx$ (Fig. 9)

$= AB \cdot$ average height.

The student should be very careful in averaging quantities to first find the actual values to be averaged. For example, the

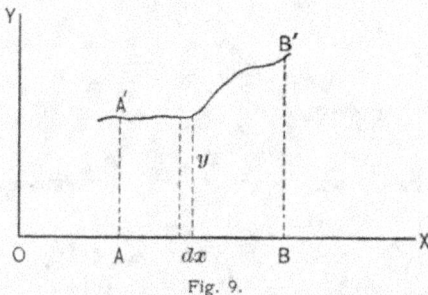

Fig. 9.

average of a series of fractions is not the average of the numerators divided by the average of the denominators. The average of a series of quantities each the product of two factors will not be the product of the average value of each factor.

V.   APPROXIMATIONS

The computation of results from physical data is often laborious, on account of the number of decimal places involved in the constants required.   In many cases, however, we may diminish the work by the use of suitable methods and approximate formulæ.   Not only is the labor of computation increased by the retention of too many decimal places, but the results so obtained are actually misleading, in that they give an appearance of accuracy not warranted by the data.   For example, any result obtained by data accurate to one part in one hundred will not be accurate to any higher degree.

Suppose that two sides of a rectangle have been measured by a metre bar divided to hundredths, and that the tenths of a division have been estimated, giving $4.258\pm$ and $6.543\pm$.   The last figure in each case is only approximate, and if the area is computed the result contains six decimal places, only three of which should in any case be retained.   The labor of writing these superfluous figures may be easily avoided by using only those partial products giving the orders we wish to retain.   We see that 4 units $\times$ .003 gives a product which we require, while .2 $\times$ .003 is of secondary importance.   The lowest partial products required are readily seen from the diagram, in which we "step down" one in the multiplicand as we "step up" one in the multiplier (Fig. 10), the arrows connecting the factors of the products required.

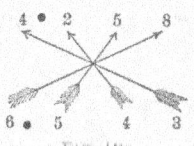

Fig. 10.

The simplest arrangement of work is that given in text-books of advanced arithmetic, and may be stated as a rule thus: *Write the multiplier in reverse order, placing the* units' *figure under the figure of the multiplicand of the same order as that to*

D   33

be retained in the product. Multiply each figure of the multiplier into the figure of the multiplicand next to the right above, and "carry" the nearest 10; then proceed as in ordinary multiplication, only writing the initial figure of each partial product in the same column, which is of the lowest order in the product.

EXAMPLE. —

$$4.258$$
$$3.456$$

$$25548$$ [Multiply by 6 as usual.
$$2129$$ [Multiply 8 by 5 and carry 4, then proceed as usual, placing 9 under 8.
$$170$$ [4 × 5, carry 2. 0 under 9.
$$13$$ [3 × 2, carry 1, etc.

$$27.860$$

Multiply 85.39738 by 1.00295, retaining four decimal places.

$$85.39738$$
$$59200.1$$

$$853974$$ [8 × 1, carry 1, etc.
$$1708$$
$$768$$
$$43$$

$$85.6493 \ Ans.$$

Many examples of this nature occur in connection with approximate formulæ, expansion coefficients, etc. The student should perform several multiplications by each method, and observe carefully the details of the shorter process.

Expressions of the form $(1 \pm \alpha)$, where α is a small quantity, are of frequent occurrence in physics. Whenever any power of such an expression is used as a multiplier or divisor, an approximate multiplier can be found by means of the "binomial theorem."

Since $[1 \pm \alpha]^n = 1 + n(\pm \alpha) + \dfrac{n(n-1)}{1 \cdot 2} \alpha^2 \cdots$ for all values of n, whether positive, negative, integral, or fractional, and, when α is small in comparison with unity, we may neglect α^2 and all higher powers of a, the approximate multiplier consists of $1 \pm na$.

EXAMPLE. — The edge of a wrought-iron cube is 20 cm. at 0° C. What will be its volume at 15° C., the coefficient of linear expansion being .0000122?

The length of each edge at 15° is

$$L_{15} = 20[1 + 15 \cdot .0000122]$$
$$= 20[1 + .000182].$$

Whence volume at $15° = 20^3[1 + .000182]^3$
$$= 20^3[1 + 3 \cdot .000182$$
$$+ \text{Higher powers of small quantities.}]$$
$$\doteq 20^3[1.000546]$$
$$= V_0[1.000546].$$

Had the volume at 24° C. been given and the volume at 0° been required, we have, in like manner,

$$V_t = V_0[1 + at]^3.$$

$$\therefore V_0 = V_t \frac{1}{(1 + at)^3} = V_t[1 + at]^{-3}$$
$$= V_t[1 - 3 at]$$
$$= V_{24}[1 - 3 \cdot 24 \cdot .0000122]$$
$$= V_{24}[1 - .0008784]$$
$$= V_{24}[.9991216].$$

When V_{24} is given, the approximate method of multiplication gives the result easily. It is to be observed that when the original length or volume is large, *i.e.* when the multiplicand is large, more decimal places in the multiplier are of importance.

As another example, consider the area of a rectangle of sides a and b when each side is slightly increased.

If a is increased by α, and b is increased by β,

$$\text{the new area} = (a + \alpha)(b + \beta)$$
$$= ab + a\beta + b\alpha + \alpha\beta \qquad \text{(Fig. 11)}$$
$$= ab + a\beta + b\alpha,$$

when $\alpha\beta$ can be neglected; *i.e.* when the corner rectangle is very small in comparison with those on the sides.

The student will be able to form approximate formulæ similar

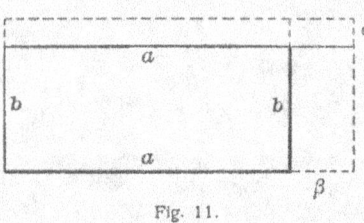

Fig. 11.

to those given in many cases, and these, in connection with the various tables, will greatly reduce tiresome numerical computations which in themselves give no insight into physical laws and phenomena.

In addition to these, a few points in connection with arrangement of work and notation may be useful.

It is customary and convenient in expressing very large or very small numbers to write only the few figures actually observed or derived, and to indicate their position by a power of 10 used as a multiplier; as,

$$45630000000 = 456.3 \cdot 10^8,$$
$$.0000122 \quad = 122 \cdot 10^{-7}, \text{ etc.}$$

In every case where numerical work is required, spend a little time and thought in a general survey of the problem.

Note in what order it is best to perform the various parts, whether factors can be cancelled or approximate values used. It is often best to write out the entire expression before any numerical work is done. Bear in mind that the understanding of the method and the facts involved is of primary importance, and numerical results are often only secondary.

MECHANICS

VELOCITY, ACCELERATION, AND FORCE

23. Express a velocity of 22 mi. per hour in (*a*) feet per minute, (*b*) kilometers per hour, (*c*) centimeters per second.

24. An express train leaves Albany at 10.13 A.M., and arrives in Buffalo at 4.45 P.M. The distance is 297 miles. Find the average velocity of the train over this distance.

25. Using velocities as ordinates and times as abscissas, draw a curve which might show the changes in velocity between any chosen time limits in a train's run. What is represented by the area included between the curve and the x-axis? What by the steepness (pitch) of the curve at any point?

26. Which is the greater velocity, 40 mi. per hour or 12 m. per second?

27. A railway train reaches a speed of a mile a minute. What is the value of this speed in kilometers per hour?

28. Speaking of the time required for light from the sun to reach the earth, Lodge says:* "If the information came by express train it would be three hundred years behind date, and the sun might have gone out in the reign of Queen Anne without our being as yet any the wiser." Verify this and compute the time which is actually required for light to reach us from the sun. (Mean distance to sun $928 \cdot 10^6$ miles.)

* *Pioneers of Science.*

37

29. The side of a cube increases at the uniform rate of 10 cm. per second.   After 2 sec. at what rate is the area of one side increasing? the volume?

30. A gun is fired on board a ship at sea; an echo is heard from a cliff after a lapse of 7 sec.   Find the distance of the ship from the cliff.   (Velocity of sound = 332 m. per sec.)

31. A man of height h walks along a level street away from an electric light of height b.   If the man's velocity is v miles per hour, find the velocity of the end of his shadow.

32. What is acceleration?   What are the dimensions of acceleration?   What is the C.G.S. unit of acceleration?

A particle has unit acceleration when it gains (or loses) unit velocity in unit time. The C.G.S. unit of velocity is a velocity of one centimeter per second. The corresponding unit of acceleration may therefore be called one centimeter per second per second. This is a somewhat cumbersome name, but it is conducive to clearness.

33. Show that the general expression for acceleration is

$$a = \frac{d^2l}{dt^2}.$$

Take a as constant, integrate twice, and discuss the resulting equations.

34. A body acquires in 4 sec. a velocity of 300 cm. per second.   What is the value of its acceleration?

$\frac{300}{4} = 75$ cm. per second per second.

35. A train having a speed of 64 km. per hour is brought to rest under the action of brakes in a distance of 510 m. What is the acceleration, if assumed to be constant?

36. What is the final speed of a body which, moving with a uniformly accelerated motion, covers 72 m. in 2 min., if

(a) the initial speed = 0,

(b) the initial speed = 15 cm. per second.

37. Plot a curve showing the relation between *distance passed over* and *time* in the case of a body having a constant acceleration. What is shown by the *pitch* of such a curve at any given point?

38. Find the distance passed over in the *t*th second by a body having a uniformly accelerated motion.

We have

space described in *t* seconds $= \frac{1}{2} at^2$,

space described in $t - 1$ seconds $= \frac{1}{2} a(t - 1)^2$;

whence space described in the *t*th second

$$= \frac{1}{2} at^2 - \frac{1}{2} a(t - 1)^2$$
$$= \frac{a}{2}(2t - 1).$$

If the body has an initial velocity v_0, we have, obviously,

space passed over in the *t*th second

$$= v_0 + \frac{a}{2}(2t - 1).$$

39. What are the ratios of the spaces passed over in successive seconds by a body moving with a constant acceleration?

40. If a body starting from rest has an acceleration of 36 cm. per second per second, over what distance will it pass in the seventh second?

41. A body has a uniform acceleration of 36 cm. per second per second. Initial velocity = o.

(*a*) How far does it travel in 8 sec.?

(*b*) How far does it travel *during* the eighth second?

42. With an initial velocity of 14 cm. per second, how answer the preceding problem?

43. A train acquires 8 min. after starting a velocity of 64 km. per hour. Assuming constant acceleration, what is the distance passed over in the fifth minute?

44. A body starting from rest with a constant acceleration passes over 18 km. the fourth hour. Find the acceleration.

$$l_{4th} = 18 = \frac{a(2 \times 4 - 1)}{2},$$

$$\tfrac{7}{2}a = 18,$$

$$a = \tfrac{36}{7} \text{ km. per hour per hour.}$$

45. A body starts from rest with a uniformly accelerated motion. In what second does it describe five times the distance described in the second second?

46. A and B are initially at the same point. If A move to the right with a uniform velocity of 6 km. per hour, and B move to the left with a uniform acceleration of 3 km. per hour per hour, what is the distance between them at the end of 4 hr.?

47. Suppose in the preceding problem that at the expiration of the 4 hr. A turns and follows B with a uniform acceleration of 4 km. per hour per hour, how long before A overtakes B?

48. A body moving with uniform acceleration passes over distances of 13 and 23 m. in the seventh and twelfth minutes respectively. Find its initial velocity and its acceleration.

49. A body starting from rest passes over 1.2 m. in the first second. The acceleration being uniform and the initial velocity zero, how long has it been in motion when it has acquired a velocity such that 6 m. are described in the last second of its motion?

50. A body m has an acceleration of 40 cm. per second per second; a body n has an acceleration of 56 cm. per second. Provided both bodies start from the same origin at the same instant and travel (a) in the same direction, (b) in opposite directions, how long before they will be 6 m. apart?

51. What definition of *force* is implied in Newton's first law? What *quantitative* definition of force is embodied in Newton's second law?

52. Discuss Newton's third law, giving one or more familiar examples.

53. Define the C.G.S. unit of force, the *dyne*.

54. Define the dyne in terms of momentum and time.

55. What is the character of the motion produced by a constant force acting on a given mass?

56. What constant force will give to a mass of 40 g. a velocity of 4.8 m. per sec. in 12 sec. ?

57. A force of 30 dynes acts on a mass of 2 g. Find the velocity acquired in 1 sec. :

$$30 = 2\,a,$$
$$a = 15.$$

Find the velocity acquired in 6 sec. :

$$v = at = 6 \times 15 = 90 \text{ cm. per sec.}$$

58. Explain fully the difference between mass and weight.

59. A body of 6 g. mass is moving with a velocity of 3.6 km. per hour. Find the force in dynes that will bring it to rest in 5 sec.

The application of a constant force to the body will produce a constant (negative) acceleration. Since the body is to lose all of its velocity in 5 sec., the rate of change of velocity, *i.e.* the acceleration is

$$a = \frac{3.6 \times 10^5}{36 \times 10^2 \times 5}$$
$$= 20.$$

And the force necessary to produce this acceleration is

$$F = ma = 6 \times 20$$
$$= 120 \text{ dynes.}$$

60. A mass of 500 g. moving at the rate of 10 m. per second is opposed by a force of 1000 dynes. How long must this force act in order to bring the body to rest ?

61. A mass of 4 kg. falls freely. What is the value of the force acting upon it ?

The acceleration due to gravity is 980 cm. per second per second. We have

$$F = Ma$$
$$= 4000 \times 980$$
$$= 392 \times 10^4 \text{ dynes.}$$

62. Show that the dyne is, roughly speaking, the weight of 1 mg., and that the unit of force in the F.P.S. system (called the *poundal*) is the weight of $\frac{1}{2}$ oz. approximately.

63. Engineers use the *weight of a pound* as the unit of force. Taking *g* as 32.2, what is the value of the unit of mass in this system ?

64. Reduce a force of 2 kg. weight to dynes.

65. Find the weight in dynes of a man who gives his weight as 140 lb.

66. What is the value of "the acceleration due to gravity" in terms of (*a*) the centimeter and second, (*b*) the foot and second, (*c*) the meter and minute?

67. Would any change occur in the weight of a ball if it were carried to the center of the earth ? Imagine the ball to be in motion at the center of the earth; is the same force required to stop it in a given time as would be required under the same conditions at the surface of the earth ?

68. Aside from any possible difference in value, would there be any advantage in buying silver in Philadelphia and selling it in Berlin, provided weighings at both places were made with the same *spring balance ?* Explain your answer fully.

69. A force equal to the weight of 2 kg. acts on a mass of 40 kg. for half a minute. Find the velocity acquired, and the space passed over in this time.

70. A force equal to the weight of a kilogram acts on a body continuously for 10 sec., causing it to describe in that time a distance of 10 m. Find the mass of the body.

71. The *weight of a pound* being taken as the unit of force (the engineer's unit, called by him simply a *pound*), find the constant horizontal pull necessary to draw a block of 12 lb. weight over a frictionless horizontal table, with an acceleration of 8.05 ft. per second per second.

In the fundamental relation

$$F = Ma,$$

we have $\qquad F = 12$ and $a = 32.2$;

whence $\qquad M = \dfrac{12}{32.2}$ units of mass.

The force required is

$$F = Ma' = \frac{12}{32.2} 8.05 = 3 \text{ lbs. weight.}$$

72. How far will a body fall from rest in five sec.? What is its final velocity? What is its mean velocity during this time?

The acceleration due to gravity is sensibly constant in any one locality. Problems in falling bodies, therefore, come under the head of uniformly accelerated motion, and the same formulas apply.

73. The Washington monument is 169 m. high. In what time will a stone fall from top to bottom?

74. What velocity does a body acquire in falling through a distance of 100 m.?

75. From what height must a body fall to acquire a velocity equal to that of an express train making 96 km. per hour?

76. A stone dropped from the top of a building strikes the ground in 3 sec. What is the height of the building?

77. A pebble thrown vertically downward from the top of a tower with a velocity of 3 m. per second, strikes the earth in 4 sec. What is the height of the tower?

78. Show that if two bodies A and B be let fall a time interval of θ apart, $A's$ velocity relative to B is constant.

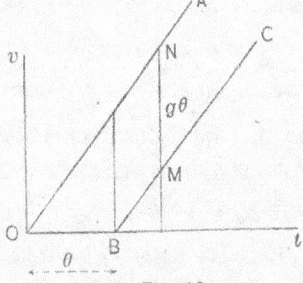

Fig. 12.

After a time t, A has acquired the velocity

$$v_A = gt;$$

But B has now been falling a time $t - \theta$ and has acquired the velocity

$$v_B = g(t - \theta).$$

Their relative velocity is therefore

$$v_A - v_B = g\theta,$$

that is, simply the velocity acquired by A before B was allowed to fall. Graphically A's velocity is represented by the line $\overline{OA}$ drawn at a pitch g; B's velocity is represented by $\overline{BC}$ drawn at the same pitch but having an intercept on the x-axis of $+\theta$. The constant intercept $\overline{MN}$ represents their relative velocity.

79. Extend the foregoing problem to the case in which both A and B have initial velocities, and discuss the conditions under which their relative velocity may be $+$, o, or $-$.

80. A body is thrown vertically upward with a velocity v_0. Find an expression for its velocity at any time t.

The student should here remember that the conditions differ from those of a body thrown downward with an initial velocity only in the direction of this velocity. In time t the body acquires the velocity gt irrespective of its initial velocity. If we count velocity upward as positive, we must have then

$$v = v_0 - gt.$$

81. A body is projected upward with a velocity of 30 m. per second. Find its velocity after 2 sec. ; after 4 sec.

82. A body is projected upward with a velocity v_0. When will it reach a given height h?

The equation of this motion is

$$h = v_0 t - \tfrac{1}{2} g t^2.$$

Its solution gives two roots which, if real, are both positive. The smaller root is the time required to reach a height h during the ascent. The greater one is the time required to reach the same height during the descent. If the roots are imaginary, v_0 is not great enough to carry the body to the height h. The student will readily interpret the case in which the roots are equal.

83. A body is projected vertically upward with a velocity of 24 m. per second. When will it reach a height of 10 m. ?

84. Show that when a body is thrown upward it has, at a height h, numerically the same velocity, whether the body be rising or falling.

85. A body is projected upward with a velocity of 20 m. per second. How high will it rise before beginning to descend?

86. A ball is thrown upward with a velocity of 20 m. per second. How long before it will cease to rise? How long before it returns to the hand?

87. The velocity of a body varies as the square of the time. If in 2 seconds after starting it has acquired a velocity of 40 cm. per second, how far will it go in 5 sec. ?

88. The velocity of a particle varies as its distance from the starting-point. Find the distance traversed in time t. Velocity at starting-point given as v_0.

NOTE. — In the following problems on the inclined plane friction is not considered; that is, the plane is assumed to be perfectly smooth.

89. Explain how the acceleration due to gravity may be studied by means of a body sliding down an inclined plane. Show that the body's acceleration along the surface of the plane varies as the vertical height of the plane. Discuss the limiting cases of this relation.

DEFINITIONS. — The *pitch* of an inclined plane is the ratio of its height to its base, *i.e.* pitch $= \frac{h}{b}$. Or, again, the pitch of a plane is the tangent of its inclination to the horizontal, *i.e.* pitch $= \tan \phi$.

In connection with roads the word *grade* is commonly used by engineers to denote the relation of the height of an incline to its length, *i.e.* grade $= \frac{h}{l}$. A "3 per cent grade," for example, means that

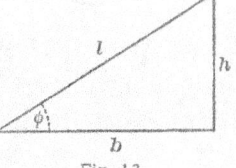

Fig. 13.

$$\frac{h}{l} = .03.$$

Obviously, *grade* $= \sin \phi$.

90. The pitch of a plane is .75. With what acceleration would a body slide down its surface?

$$a = g \sin \alpha = 980 \cdot \tfrac{3}{5} = 588.$$

91. Which is the steeper, a 6 per cent *grade* or a 6 per cent *pitch?*

92. A body sliding down an inclined plane describes in the third second of its motion a distance of 122.5 cm. Find the grade.

$$l = \frac{a(2t-1)}{2},$$

$$a = \frac{2 \times 122.5}{5} = 49 \text{ cm. per second per second}$$

$$\text{Grade} = \frac{a}{g} = \frac{49}{980} = \frac{1}{20} = 5 \text{ per cent.}$$

93. A body slides down the plane *OA*. Show that the velocity acquired on reaching *A* is the same as that which would be acquired in a free fall through the distance *OH*.

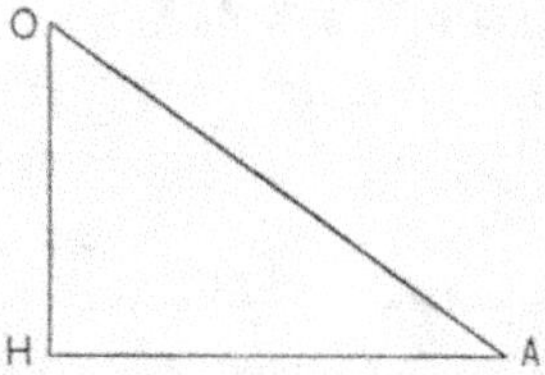

Fig. 14.

94. A heavy particle slides from rest down an inclined plane whose length is 4 m. and whose height is 1.2 m. What is the velocity of the particle on reaching the ground? What is the time of fall?

95. A man can just lift 150 lb. What mass can he drag at a uniform rate up a frictionless grade of 7.5 per cent?

$$150 = x \frac{7.5}{100},$$

$$x = 2000 \text{ lb.}$$

96. A body slides down a plane 2.1 m. long in 3 sec.; to slide down another plane of the same height requires 5 sec. What is the length of the latter plane?

97. A body slides freely down an inclined plane. The distances passed over in successive seconds are in what ratio? (Compare with 40.)

98. A board is 4.95 m. long. To what angle must it be tipped in order that a body shall slide the full length in 3 sec.?

99. The height of an inclined plane is 426 cm. and its grade is 30 per cent. With what initial velocity must a particle be projected upward along the plane in order to come to rest just at the summit?

100. A number of planes have lengths and inclinations equal to the chords *OA*, *OB*, etc. Show that if a number of particles are allowed to slide down these planes, all starting from *O* at the same instant and without initial velocity, they will all reach the circumference in the same time.

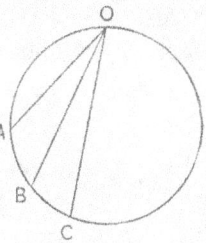

Fig. 15.

101. A point and a line lie in a vertical plane. Find the line of quickest descent from the point to the line.

102. A freight train is moving at the rate of 8 mi. per hour; a train man runs over the cars towards the rear of the train, a distance of 220 ft., in 30 sec. What is his speed relative to the surface of the earth?

103. Two trains of the same length are running with the same velocity on parallel tracks, but in opposite directions. Their combined length is 800 ft., and they pass each other in 6 sec. What is the velocity of the trains relative to the track?

104. A and B are at one corner of a square. They desire to reach the diagonally opposite corner at the same instant. A chooses the diagonal path, while B follows around two sides. (*a*) What ratio must exist between the *magnitudes* of their velocities? (It is assumed that these magnitudes are constant.)

105. The component of a ship's velocity in an easterly direction is 7.2 mi. per hour; the component in a southerly direction is 4.6 mi. per hour. What is the total velocity of the ship? What is its direction of motion?

106. When a ship is sailing northeast at the rate of 10 mi. per hour, with what speed is it approaching a north and south coast lying to the east?

107. A steamer is moving due north with a velocity of 25.6 km. per hour. The smoke from the funnel lies 35° south of east. If the wind is due west, find its velocity.

108. A body is moving upward along a path inclined 30° to the horizontal with a velocity of 60 m. per minute. (*a*) What is its velocity in a horizontal direction, (*b*) in a vertical direction, (*c*) at right angles to the direction of motion?

109. A street car is moving at the uniform rate of 6 mi. per hour up a 5 per cent grade. Find the velocity in feet per minute with which the car is rising vertically.

110. Find the resultant of the velocities 8 and 10 m. per second when the angle between them is 30°, 45°, 150°, and 180°.

111. Given four velocities *a*, *b*, *c*, and *d* of magnitudes 6, 8, 12, and 20 units respectively. The angle between *a* and *b* is 30°, that between *b* and *c* is 15°, and that between *c* and *d* is 80°. Find by resolving these velocities along any two rectangular axes their resultant in direction and magnitude. (See Introduction.)

112. A man starts to row across a stream at a velocity of 4.4 mi. per hour. If the velocity of the current at all points be 3 mi. per hour, at what angle to either bank must he make his course in order to land at a point directly opposite that from which he started? If there were no current, at what speed should he row directly across in order to make the trip in the same time as under the foregoing conditions?

113. A point is moving along a straight line with an acceleration of 22 cm. per second per second. Find the acceleration of the point in directions 30°, 90°, and 180° from this line.

114. A particle is projected upward at an angle of 45° to the horizontal with a velocity of 120 m. per second. In what time will it reach its greatest height?

SUGGESTION. — When the body reaches its greatest height, the vertical component of its velocity must be zero. Hence find the vertical component of the initial velocity, and divide by the *loss* of velocity per second; that is, find the time required for the body to lose *all* of its initial velocity in a vertical direction.

115. A particle is projected upward at an angle of 30° to the horizontal with a velocity of 70 m. per second. Find the *time of flight*, *i.e.* the time elapsing before the particle again reaches the horizontal.

116. A body is projected with a velocity V at an angle a. Find the horizontal distance (the *range*) described.

Without considering the nature of the path, the range is readily obtained by multiplying the horizontal velocity, which is constant, by the time of flight.

117. For a given initial velocity, show that the range is a maximum when the body is projected at an angle of 45°.

118. A body is projected at a given angle a to the horizontal. If the initial velocity be doubled, how does the range vary?

119. Show that any two complementary angles of projection give the same range.

120. Find the greatest height to which a body will rise and its range, if it is projected with horizontal and vertical velocities of 40 and 80 m. per second.

121. A body is thrown horizontally from the top of a tower 100 ft. high with a velocity of 200 ft. per second. Find

(*a*) the time of flight,

(*b*) the range,

(*c*) the velocity with which the body strikes the ground,

(*d*) the angle at which it strikes the horizontal.

122. Find the equation of the path of a projectile, and show that the trajectory is a parabola.

123. Find an expression for the angle at which a particle must be projected with a velocity of given magnitude in order that it shall pass through a given point in the plane of the motion. What indicates that the given point is *out of range?*

124. (*a*) Define *angular velocity*. (*b*) Find the angular velocity of a wheel making 1000 revolutions per minute.

In engineering practice it is common to express rate of rotation in revolutions per minute. In these units the angular velocity would be simply 1000. But in physics the velocity would be taken in *radians per second*.

E

125. Compare the angular and linear velocities of two points distant 1 and 2 m. respectively from the center of a wheel making 40 revolutions per minute.

126. What are the dimensions of angular velocity?

127. A wheel makes 1 revolution in .5 sec. What is its angular velocity?

128. Express the angular velocity of the rotation of the earth on its axis in radians per second.

$$\frac{2\pi}{24 \times 3600} \text{ radians per second.}$$

129. What is the linear velocity of a point on the surface of earth at 60° north latitude? (Rotation alone considered. Mean radius of earth 6366.8 km.)

130. A pinion having 16 teeth is geared to another having 66 teeth. Compare the angular velocities.

131. The driving wheel of a locomotive is 1.5 m. in diameter. If the wheel makes 250 revolutions per minute, what is the mean linear velocity of a point on the periphery? What is the velocity of the point when it is vertically above the axis of rotation? When it is vertically below?

132. A freely falling body acquires a momentum of 12,054 C.G.S. units in 3 sec. What is its mass?

133. The velocities of two bodies are as 6 : 4, and their momenta are as 9 : 2. What is the ratio of their masses?

$$\frac{6}{4} \cdot \frac{m}{m'} = \frac{9}{2};$$

$$\frac{m}{m'} = \frac{36}{12} = \frac{3}{1}.$$

134. The mass of a gun is 4 tons. If a shot of mass 20 lb. be fired with an initial velocity of 1000 ft. per second, what is the initial velocity of the recoil?

135. What pressure will a man weighing 150 lb. exert on the floor of an elevator descending with an acceleration of 4 ft. per sec. per sec. ? Explain the sensation of being lifted which one has in an elevator suddenly arrested in its descent.

136. A balloon rises with a uniform acceleration of 4 m. per second per second, carrying with it a spring balance upon the hook of which is hung a ball of 7.35 kg. weight. (*a*) What is the reading of the balance in kilograms' weight? (*b*) What reading would the balance show if the balloon were *descending* with the acceleration named?

137. Two masses M and m are connected by an inextensible string passing over a smooth peg. Neglecting the mass of the string, find: (*a*) the acceleration of M and m, and (*b*) the tension of the string.

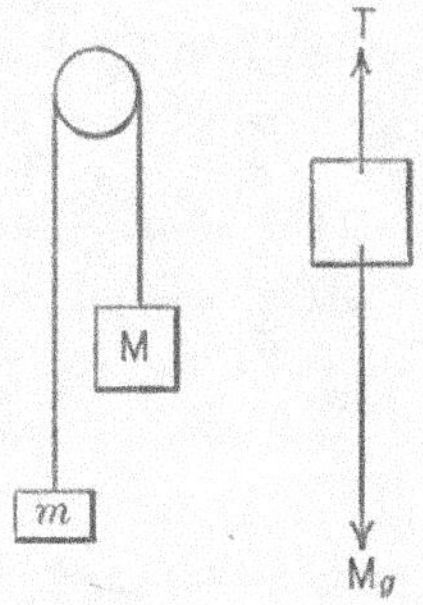

Since the string is without mass, and since it does not stretch, it has the same tension T at every point in its length. Further, the downward velocity of M must equal the upward velocity of m, and their accelerations must be numerically equal. Let a be this common value.

Consider the forces acting on M. These are: (1) the weight of M downwards, and (2) the tension T upwards. And *there are no others*. Hence we write

$$Mg - T = Ma.$$

Fig. 16.

Again, considering the forces acting on m, we arrive at a similar relation, and, from the two equations thus found, the values of a and T are readily deduced.

138. Show that the value of a found above is independent of the unit in which M and m are measured. Can the same be proved of T?

139. If the masses M and m are equal, what kind of motion is possible? What is the value of the tension T?

140. Two masses are connected by a weightless cord hanging over a smooth peg; the sum of the masses is twice their difference. Find the common acceleration.

141. Show that, in order to derive the expression for the acceleration in 137, it is not necessary to consider the tension in the cord.

142. A cord passing over a frictionless pulley has fastened to its ends masses of 5 and 10 kg. respectively. Find the pull on the hook sustaining the pulley when the masses are in motion. (Neglect weight of pulley itself.)

143. Explain how the value of g may be determined by Atwood's machine.

144. One has weights aggregating 10 kg.; it is required to divide the total into two parts such that when connected by a string passing over a pulley, the whole will have an acceleration $\frac{1}{8}$ that due to a free fall.

145. A mass m is drawn horizontally along a smooth table by a cord passing over a small frictionless pulley and attached to a mass M. Find expressions for the acceleration of both masses and the tension in the cord.

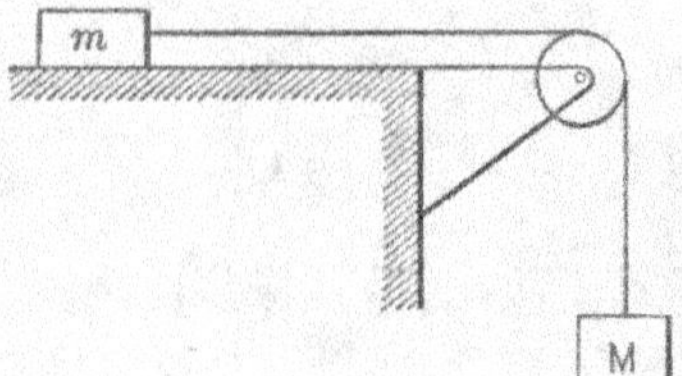

Fig. 17.

146. In the last problem what must be the ratio of M to m in order to produce an acceleration equal to $\frac{2}{5}$ that of a freely falling body?

147. A mass of 20 g. hanging over the edge of a table draws a mass of 84 g. along the horizontal surface. Assuming no friction, find the tension in the cord. In what time will the second mass traverse the length of the table if this latter is 3 m. long?

148. Two masses m_1 and m_2 are connected by a string. m_1 hangs freely while m_2 rests on a plane inclined at an angle α to the horizontal. If the string passes over a small frictionless pulley at the summit of the plane, find the resulting acceleration.

Consider the forces acting on m_1. These are: (1) its weight m_1g and (2) the cord tension T. If f be the common acceleration, we must have

$$m_1g - T = m_1f.$$

So, the forces acting on m_2 are the resolved part of its weight acting along the

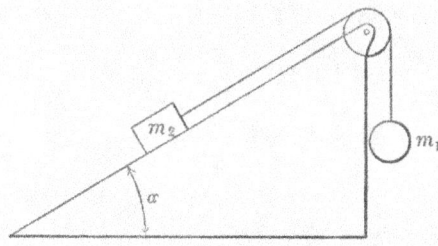

Fig. 18.

plane and the cord tension. This gives another and similar equation in which f and T are unknown. By eliminating these quantities are readily found.

149. Show that when $\alpha = 90°$, the results are identical with those obtained in 142; also that when $\alpha = 0$, the results are identical with those in 145.

150. In order to pull a mass of 1000 kg. up an incline of 30°, a rope and pulley are used as in 148. Neglecting all friction, compute the tension in the rope when a mass is used sufficient to cause an acceleration of 0.4 m. per second per second.

151. Find the resultant of two forces of 6 and 9 kg. weight:
(1) Acting in the same straight line and in the same direction.
(2) Acting in the same straight line but in opposite directions.
(3) Acting at angles of 30°, 45°, 90°, 120°, and 150°.

152. A force is inclined 36° to the horizontal. What is the ratio of its vertical to its horizontal component?

153. Three concurrent forces of 8, 30, and 12 kg. weight are inclined to the horizontal by angles of 32, 60, and 143° respectively. Find the horizontal and vertical components of their resultant.

154. Two forces acting at an angle of 60° have a resultant equal to $2\sqrt{3}$ dynes. If one of the forces be 2 dynes, find the other force.

155. Two equal forces act on a particle. If the square of their resultant is equal to three times their product, what is the angle between the forces?

156. At what angle must two forces act so that their resultant may equal each of them?

157. Find the angle θ which shall make the resultant of two forces of constant magnitude a maximum.

158. Let the angle between two forces of constant magnitude increase continuously from o to π. Discuss the variation of the angle between the resultant and one of the forces.

159. Show that when three forces in the same plane are in equilibrium their lines of action meet in a point.

160. Show that when three forces are in equilibrium each force is proportional to the sine of the angle between the other two (Lami's theorem).

161. Find by graphic construction the resultant of four forces of 3, 7, 5, and 12 lb. weight acting on a particle, and represented in direction by the successive sides of a square.

162. Two forces of 3 and 4 units respectively are balanced by a third force of $\sqrt{37}$ units. Find the angle between the first two forces.

163. A mass of 4 kg. is suspended at the middle of a cord whose two halves make an angle of 30° with the horizontal. What is the tension in the cord? (Mass of cord neglected.) The mass remaining the same, how may the tension in the cord be varied? Discuss the law of variation.

164. A weight of 14 kg. hangs at the end of a string; a force is applied horizontally deflecting the string 30° from the vertical. What is the value of this force and what the tension in the string?

165. A string connecting two equal masses hangs over three smooth, equidistant pegs. Neglecting the weight of the string, find the resultant pressure on each peg.

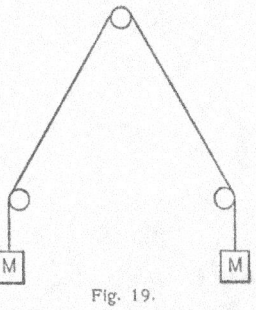

Fig. 19.

166. Why is a long line desirable in towing a canal boat? To pull a canal boat at a uniform rate requires a force in the direction of motion of P lb. weight. If the rope make an angle of 10° with the line of motion, and if the weight of the rope be neglected, what pull must the horses exert?

167. A body of weight 30 kg. is suspended by two strings of lengths 5 and 12 m., attached to two points in the same horizontal line whose distance apart is 13 m. Find the tensions in the strings.

168. A mass of 40 g. rests on a plane inclined at 30°. Find in grams' weight the force parallel to the plane: (1) necessary to hold it there, (2) necessary to draw it uniformly up the plane, (3) necessary to cause an acceleration of 30 cm. per second per second up the plane.

169. A block having a mass of 100 g. is prevented from sliding down an inclined plane by means of a cleat. Find the inclination of the plane which will make the pressure on the plane equal that on the cleat, and give the numerical value of their sum.

170. A block is held from sliding down an inclined plane by a cleat. Plot two curves showing the variations of the pressure exerted by the block (1) on the plane and (2) on the cleat, with variations of the angle of the plane.

171. Determine analytically the angle for which the sum of the cleat pressure and plane pressure is a maximum.

172. A ball is held at rest on an inclined plane *of given angle* α by means of a cord. Find the cord tension when the angle between the cord and plane is θ. For what value of θ is this tension a minimum?

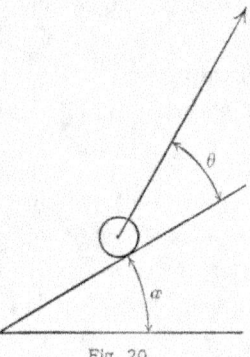

Fig. 20.

173. The upper end of a ladder rests against a smooth vertical wall; the lower end on a smooth horizontal floor, slipping being prevented by means of a cleat. The ladder is of uniform cross-section, weighs 100 lb., and is inclined at 60° to the horizontal. Find the reactions of the different surfaces against which the ladder rests.

174. When a person sits in a hammock the tension on either sustaining hook is greater than the person's weight. Explain. Does the tension increase or decrease as the hammock is made more nearly horizontal?

175. A string hanging over a pulley has at one end a mass of 10 kg. and at the other masses of 8 kg. and 4 kg. When the system has been in motion for 5 sec., the 4 kg. mass is removed. Find how much farther the weights go before coming to rest.

176. The ram of a pile driver weighs 500 lb. It is allowed to fall 20 ft. driving a pile 6 in. Find the value of the resistance, assuming it to be uniform.

[Consider the acceleration needed to bring the body to rest in the given distance.]

177. Show graphically how to find the resultant of two parallel forces, (a) when the forces are *like*, and (b) when the forces are *unlike*.

178. Apply the graphical construction to the case of two equal, *unlike* forces and interpret the result.

179. A man carries a bundle at the end of a stick placed over his shoulder. If he varies the distance between his hand and his shoulder, how does the pressure on his shoulder change?

180. The resultant of two like parallel forces is 16 kg. weight and its point of application is 6 cm. from that of the larger force, which is 10 kg. weight. Find the distance of the smaller force from the resultant.

181. Equal weights hang from the corners of a triangle which is itself without weight. Find the point at which the triangle must be supported in order to lie horizontally.

SUGGESTION. — The forces at the corners are all equal and parallel. The resultant of any two must act at the mid-point of the side connecting them. Combine this partial resultant with the force at the third corner.

182. A teamster considers one horse of his pair as 25 per cent stronger than the other. At what point should the bolt be placed in the "evener" in order that each horse may draw in proportion to his strength?

183. A bridge girder rests on two piers distant a feet apart. The girder is of uniform cross-section, p lb. weight per linear foot. At a distance $\frac{2}{5}a$ from one end a load of P lb. weight is placed. Find the reactions of the piers.

184. What is a couple and what is the moment of a couple?

185. Show that the algebraic sum of the moments of the two forces forming a couple about any point in their plane is constant.

186. One of the forces of a couple is 60 dynes; the distance between the forces is 0.3 m. Find the moment of the couple.

187. A straight bar is acted upon at its ends by two equal and parallel but opposite forces of 12 kg. weight each. The bar makes an angle of 45° with the direction of the forces and is 3 m. long. Find the moment of the resulting couple.

CENTER OF INERTIA (OR MASS) (OR GRAVITY)

188. Two equal weights are connected by a light, stiff rod. Find the center of inertia.

189. How would the center of inertia be moved if one of the weights were doubled? if both were multiplied by three?

190. Three weights, 4, 5, and 7, are joined by stiff weightless rods. Find the center of mass of the system.

191. What is the center of gravity of a triangle? a square? a parallelogram? a trapezoid? Test your answers with pieces of cardboard.

192. The diagonals of a square are drawn, and one of the triangles resulting is removed. Find the center of gravity of the remaining figure.

193. Two lines are found on a surface such that the surface will "balance" about each. What point is determined by their intersection?

194. Four masses are supposed concentrated at the points A, B, C, D; masses 9, 5, 6, 10, respectively. The lengths OA, AB, BC, CD are 5, 8, 4, 10, respectively. Find the distance of the center of mass of the system from the point O.

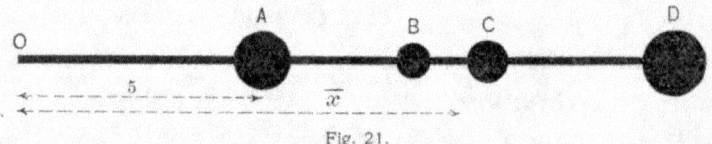
Fig. 21.

We have $5 \cdot 9 + 13 \cdot 5 + 17 \cdot 6 + 27 \cdot 10 = $ sum of mass-distance products,
$$9 + 5 + 6 + 10 = \text{sum of masses}.$$

$\therefore$ distance required is $\frac{492}{30} = 16+$.

The distance from O to the center of gravity may be found from an equation expressing the fact that about that point the sum of the moments of the couples due to gravity is zero.

Let $\bar{x}$ = distance required.

Then lever arm for gravity action on A is $\bar{x} - 5$.

Whence moment of couple due to A is $g\,(\bar{x} - 5)9$,

couple due to B is $g\,(\bar{x} - 13)5$,

couple due to C is $g\,(\bar{x} - 17)6$,

couple due to D is $g\,(\bar{x} - 27)10$.

Sum equals o. $\therefore$ $30\,\bar{x} = 482$, $\bar{x} = 16+$, as before.

195. A body is suspended by a flexible cord. What position will the center of gravity assume? Explain.

196. Explain the connection between the center of gravity of a body and its stability.

197. Express the fact of no resultant couple about the center of gravity in the notation of the calculus.

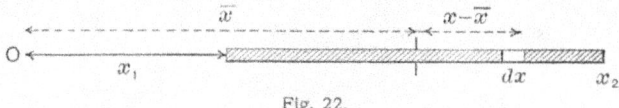

Fig. 22.

When the body is linear or is symmetrical about a line.

Let $\qquad\qquad \bar{x}$ = the distance of C.G. from O,

x = the distance of any mass element from O,

dx = the length of element.

Then $\qquad\qquad \rho dx$ = mass element,

$x - \bar{x}$ = lever arm.

$\therefore$ mom. of couple $= \rho dx\,(x - \bar{x})\,g$.

Sum of mom. $= \displaystyle\int_{x_1}^{x_2} \rho dx\,(x - \bar{x}) = 0$. [By def. of C.G.

$$\therefore \bar{x} = \frac{\displaystyle\int_{x_1}^{x_2} \rho x\,dx}{\displaystyle\int_{x_1}^{x_2} \rho\,dx}.$$

(a) Find $\bar{x}$ for a *uniform* rod of length l.

(b) Find $\bar{x}$ for a rod where ρ increases from x_1 to x_2, *i.e.* where $\rho_x = k \cdot x + \rho_0$.

(c) Find $\bar{x}$ for an isosceles triangle.

198. Show directly from the definition of C.G. that its co-ordinates are given by three equations of the form $\bar{x} = \dfrac{\int \rho x \, dv}{\int \rho \, dv}$.

199. Explain the distinction in meaning and use between the above expression for $\bar{x}$ and $x = \dfrac{\Sigma m x}{\Sigma m}$.

200. Find C.G. of a cone of revolution.

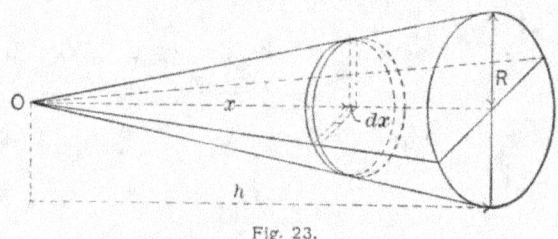

Fig. 23.

Take dv as a slice ∥ to base. Then

$$dv = \frac{R^2}{h^2} \cdot \pi x^2 dx \, ;$$

$$x = \frac{\int_0^h \pi \rho \frac{R^2}{h^2} x^3 dx}{\int_0^h \pi \rho \frac{R^2}{h^2} x^2 dx} = \tfrac{3}{4} h.$$

201. Find C.G. of a sector of a circle.

202. Find C.G. of a segment of a circle.

203. Find C.G. of an arc of a circle.

204. Apply the general formula for the co-ordinates of the C.G. to the square, the circle, the rectangle, the triangle.

205. Two bodies, attracting each other with a force measured by $\dfrac{m_1 m_2}{r^2}$, move toward each other. Where will they meet?

206. Show that the momentum of any system of bodies, each of which has motion of translation only, is the same as the momentum of the sum of the masses moving with the velocity of the center of gravity of the system.

207. Two masses are joined by a rigid rod; the system is thrown in the air so that it whirls. What will be its center of rotation?

208. Two spheres glide freely on a light, rigid rod, and are joined by a spiral spring sliding freely on the rod; the system is thrown so that the rod has an initial angular velocity ω_0. Discuss the relative position of the two spheres with reference to the center of gravity of the system.

WORK AND ENERGY

209. A constant force of 20 dynes moves a body 100 cm. What work is done?

210. A force of 9000 dynes is exerted constantly on a body, and moves it 4 m. per second. How much work is done in 1 min.?

211. How much work is required to lift 1 kg. from the sea level to a height of 1 m. where $g = 980$? 3 kg.? 8 kg.?

212. How much work is required to raise 1 kg. 2 m.? 2 kg. 5 m.?

213. What work is required to raise 80 kg. 3 m. against gravity? 10 m.?

214. Raising 80 kg. 8 m. is equivalent to raising 40 kg. how many meters? To lifting what mass 5 m.?

215. $98 \cdot 10^{10}$ ergs are expended in raising 100 kg. How high were they raised?

216. A force of 40 dynes is applied at an angle of 60° to the path along which the point of application moves. What work will be done when the point is moved 1000 cm.?

217. $8 \cdot 10^8$ ergs of work are required to move a body 400 m. in a straight line. What force is required if applied at an angle of 10° with the path? of 20°? of 30°? of 80°?

218. $4 \cdot 10^8$ ergs of work are required to move a body $8 \cdot 10^4$ cm. What was the average force required?

219. $6 \cdot 10^{10}$ ergs of work have been expended in moving a body against a resisting force of $3 \cdot 10^5$ dynes. How far was it moved?

220. A stone of volume 10^3 c.c., sp. gr. 2.6, is raised from the bottom of a lake to the surface, a distance of 20 m. Find the work done. See Ex. 422.

221. Find the work done in forcing a block of wood, volume $8 \cdot 10^4$ c.c., sp. gr. .7, to the bottom of a tank of water 4 m. deep. What if tank were filled with mercury?

222. Show that if gravity be the only resisting force, the *work* done on a given mass in raising it a given height is independent of the path. Or that the force required always decreases in the same ratio as the path increases.

223. Show why it is easier to draw a load up an inclined plane than lift it vertically, neglecting friction. What element is decreased? What increased?

224. A vertical tank having its base in a horizontal plane is to be filled with water from a source in that plane. The area of the cross-section is 4 sq. m., the height is 6 m. Find the work required to fill it.

225. Show that the work required to raise a system of bodies each to a certain height is the same as the work required to raise the entire mass to a height equal to that through which the center of gravity of the system is raised.

226. A body is raised 80 m. against a force which constantly increases. The initial value of the force is 40 dynes, its final value 460 dynes. If the force increased uniformly with the distance moved, how much work was done?

227. In an ordinary swing is the force required to displace the swing constant? If not, how could the work be computed?

228. A uniform rod 10 m. long, and mass per centimeter length 5 kg., is drawn vertically upward a height of 10 m. How much work is done? How much work would be required to raise the rod from a horizontal to a vertical position?

NOTE. — Consider the *average* height of elements of mass.

229. A plank 4 m. long is hinged at one end. The plank is raised so as to make an angle of 45° with the horizontal. What work is done? (Mass of 1 cm. of plank 9 kg.)

230. Express work in terms of mass, acceleration, and distance.

231. If the unit of time were taken as 2 sec., how would the unit of work be altered?

232. Show that power = force × velocity. What does the statement mean when the velocity is changing? In what units must force and velocity be measured so that power may be expressed in ergs per second?

233. In what two general ways is the energy of a railway locomotive expended while the train is acquiring velocity?

234. The force required to overcome the friction of a wagon on a certain road is $2 \cdot 10^{10}$ dynes. How much work is done in drawing it 20 km.?

235. On a perfectly level road it was found that the pull required to keep a wagon moving uniformly was .01 of its weight. What work is done in drawing a wagon weighing 2000 kg. a distance of 3 km.?

236. A man presses a tool on a grindstone with a force equal to 10 kg. weight. The circumference of the stone is 3 m., the coefficient of friction .2. How much work is done in one turn of the crank? (Neglecting friction of bearings, etc.)

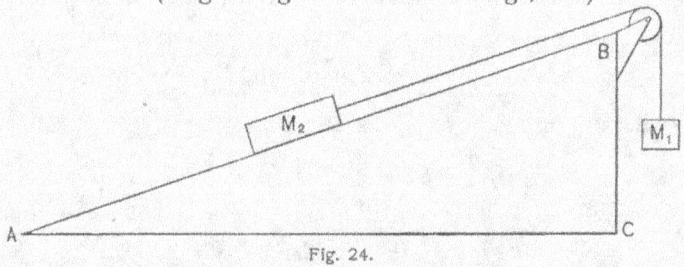

Fig. 24.

237. If $BC = .1\,AB$, what mass at M_1 will draw M_2 up AB without acceleration, neglecting friction? What effect would be observed if a greater mass were placed at M_1?

238. State how you could apply the principle of work to above case when there is friction.

239. Find the work done in drawing 120 kg. up an inclined plane of base 4 m., height 3 m., $\mu = \frac{2}{15}$.

240. How much of the work is due to friction?

241. A mass of 100 g. is moving in a circle of radius 1 m., and makes 10 revolutions per second. What is its kinetic energy? What would be its energy if the circle were half as large?

242. Five masses, 3, 8, 5, 7, and 11 g., are attached at distances 11, 7, 5, 8, 3 cm., respectively, from the centre of a wheel making 20 revolutions per second. Find the kinetic energy of each. How far from the center could the whole mass be placed so that the energy would be the same?

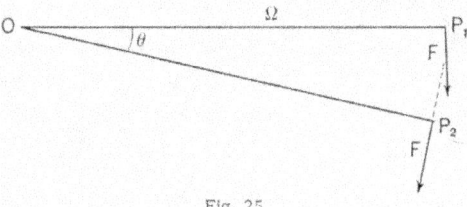

Fig. 25.

Let a constant force F be applied at a point r distant from $O \perp OP_1$. If the rod OP_1 is rigid, the work done in turning through an angle θ, since $P_1P_2 = r\theta$, is $Fr\theta = $ force × displacement. So work done by a couple or torque

> = moment of couple (Fr) × angle turned through
>
> = torque × angle turned through
>
> = average torque × angle turned through
>
> when torque is not constant
>
> $= \int Fr d\theta$. [Where $Fr = f(\theta)$.

243. A shaft s turns 120 times per minute. The radius of the shaft is 2 cm. The distance from the center of the shaft to the point where the mass is applied is 2 m. It requires a mass

F

of 80 kg. to hold the lever in equilibrium. Find the work done in 5 min.

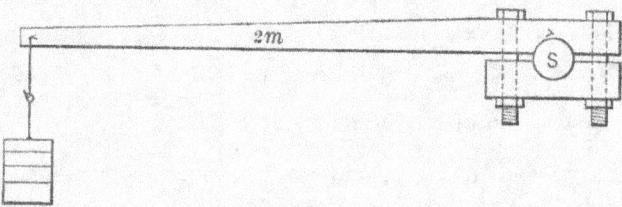

Fig. 26.

244. A mass of 80 g., moving with a velocity of 10 cm. per second, has what kinetic energy?

245. What is the kinetic energy of a bullet, mass 100 g., velocity 150 m. per second?

246. A body of mass 60 g. has a velocity 40 cm. per second, and an acceleration of 10 cm. per second per second. How much kinetic energy will it acquire in the next second? How much the fifth second later?

247. A body of mass 5 kg. is given an initial velocity of 20 m. per second on smooth ice. If the total average resisting force which it encounters is 10^5 dynes, how far will it go before coming to rest? How much energy will it have when it has gone half the distance?

248. A ball of mass 4 kg., velocity 80 m. per second, penetrates a bank of earth to a depth of 2 m. Find average resistance.

249. A ball of mass 10 g. enters a plank with a velocity of 10 m. per second and leaves it with a velocity of 2 m. per second. How much energy has it lost?

250. If the plank is 20 cm. thick and all the work is expended in piercing it, what is the average resistance?

251. A bullet is fired vertically upward with an initial velocity of 500 m. per second. What is its kinetic energy: (*a*) initially? (*b*) when half-way up? (*c*) at its highest point? (*d*) when half-way back? What is its potential energy in each case? What is the sum of E_k and E_p in each case?

252. A mass *m* falling freely acquires how much kinetic energy per centimeter of its fall? It loses how much potential energy?

253. Two balls of mass 100 and 200 kg. are attached to a firm light rod. The distance between the centers of the balls is 1 m. The system is thrown so that the center of gravity has a velocity of 20 m. per second, and the system turns ten times per second around this center. Find the kinetic energy of the system.

254. Compare their energies of rotation about the center of gravity of the system.

255. What is meant by the term "closed system" as applied to energy? Give examples.

256. State in words the relation between the work done on a system by an external force and the rate of gain of energy by the system and the losses by friction.

257. Trace the energy changes in a single vibration of a pendulum: (1) When the air resistance may be neglected. (2) When air resistance is taken into account.

258. Express in calculus notation the statement that the sum of the potential and kinetic energy of the bob of a simple pendulum is constant.

259. A mass of 60 g. is vibrating in a straight line with S.H.M. The length of the line is 4 cm., the periodic time is 2 sec. What is its average kinetic energy?

260. The velocity of a bullet is decreased from 500 to 400 m. per second by passing through an obstacle; its mass is 100 g. What energy has it lost? What has become of that energy?

261. Calculate (in ergs, and also in kilogram-meters) the work necessary to discharge a bullet weighing 10 g., with a velocity of 10,000 cm. per second.

262. If the potential energy of a stone of mass *m* and at a height *h* is entirely converted into kinetic energy, find the velocity it must acquire.　Would air friction increase or decrease this velocity?

263. If the stone were attached to a very flexible and extensible spring, what alteration of energy distribution would occur?

264. A solid sphere of cast iron is *rolling* up an incline of 30°, and at a certain instant its center has a velocity of 40 cm. per second.　Explain how to find how far it will ascend the incline, neglecting friction of all kinds.　Would the distance be the same if it were sliding up the incline?

265. If the sphere were hollow, would it acquire the same velocity as the solid one in rolling the same distance down the plane?

266. What are the dimensions of power?　If the unit of time were the minute, the unit of length the meter, how would the unit of mass need to be altered that a given power should be expressed by the same number?

267. Define erg, joule, watt.

268. A constant force is applied to a body on a horizontal plane.　If the applied force is greater than the friction between the body and the plane, why cannot an infinite velocity be obtained?

269. The mass of a car is 2000 kg.　The resistance due to friction is $12 \cdot 10^4$ dynes.　A man pushes the car with a force which would support a mass 90 kg.　His maximum power is $746 \cdot 10^6$ ergs per second.　How long can he continue to exert his full *force?*

When the component of force along the path of the point of application is variable, we must find how its magnitude varies along this path and apply the integral calculus to add up the elements of work.

·When
$$W = \int_{x_1}^{x_2} F dx,$$

where F must be expressed in terms of x, *i.e.* $F = f(x)$.

The cases of most interest are perhaps when

$$f(x) = kx,$$ [k a constant.

$$f(x) = \frac{k}{x^2}.$$

The first applies to cases of compression and stretching, as springs, etc. ; the second to gravitation, electricity, and magnetism, etc.

270. When $F = 5x$, find the work done in displacing a body 100 m.

Here $$W = \int_0^{10^4} 5\,x dx = \left[\frac{5\,x^2}{2}\right]_0^{10^4} = \tfrac{5}{2} \cdot 10^8 \text{ ergs,}$$

which is the same as taking half the sum of the initial and final force, and multiplying by entire displacement.

271. When $F = \dfrac{20}{x^2}$, find work done in displacing the point

of application from $x = 20$ to $x = 220$.

$$W = \int_{20}^{220} 20\,\frac{dx}{x^2}.$$

Could this result be obtained by taking $\tfrac{1}{2}$ (final force — initial force) × displacement?

272. A coiled spring is attached to a 50 kg. weight. What work is done if the increase of length of the spring is 2 m. when the weight is just lifted ?

273. If the pressure of a gas increases as its volume decreases, show how work done in compression could be computed.

274. A horse is hitched to a loaded wagon by a long extensible spring. Does the work done by the horse in just starting depend on the ease with which the spring is stretched ?

275. A bicycle rider moves up a grade against the wind. Against what forces does he do work ? In what ways does he expend energy ? From which of these expenditures can he get a return of energy, and how ?

The general expression for work may be written $W = \Sigma F ds$, where ds is so short that F may be considered constant over its length. We may then resolve both F and ds along any three lines we please, as OX, OY, OZ.

Let x, y, z, components of F, be X, Y, Z.
Let x, y, z, components of ds, be dx, dy, dz.

Then
$$W = \int [Xdx + Ydy + Zdz]$$
$$= \int_{s_1}^{s_2} \left[X\frac{dx}{ds} + Y\frac{dy}{ds} + Z\frac{dz}{ds} \right] ds,$$

where X, Y, Z may depend on x, y, z.

When F is constant and along s, the formula reduces to $\int Fds$, as the student may prove.

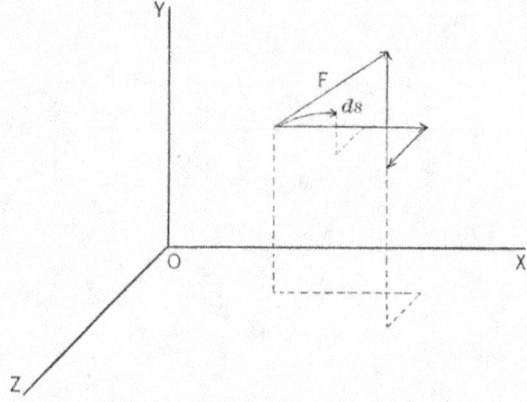

Fig. 27.

As an example we may take the work done by a couple in turning through 360°. Taking the plane of xy as the plane in which the lever arm lies, we have

$$W = \int [Xdx + Ydy].$$

By symmetry we see that the

$$\int Xdx = \int Ydy. \quad \therefore \quad W = 2\int Xdx,$$
$$X = F \sin \theta, \qquad x = r \cos \theta,$$
$$dx = - r \sin \theta d\theta,$$
$$Xdx = - Fr \sin^2 \theta d\theta,$$
$$W = 2 Fr \int_0^{2\pi} \sin^2 \theta d\theta = F \cdot 2 \pi r = F \cdot \text{circumference of } \odot.$$

It is often convenient to use the law of the conservation of energy in the solution of problems dealing with machines of various types. To do this, we form an equation involving the element required; one member of the equation representing all the work expended on the machine, the other all the work done by the machine. That is, equate the entire energy supplied to the machine to the entire energy used, stored, and wasted.

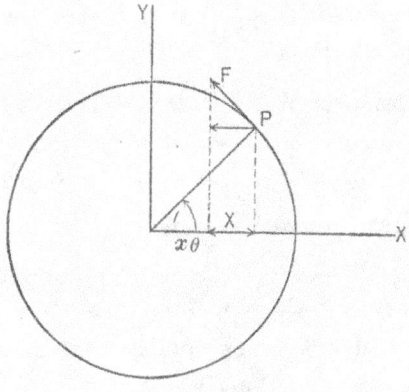

Fig. 28.

The energy given to the machine may be used in various ways; as,

(1) Lifting weights, etc. (visible and useful work).
(2) Overcoming friction (waste, transformed to heat).
(3) Strain of parts of machine (potential energy).
(4) Momentum of parts of machine (kinetic energy).
(5) Transformed to other forms, as electric, chemical, etc.

The complete analytical expression in case all of these are considered is likely to be very complicated. We therefore simplify matters by neglecting certain items of relatively small importance, yet it should be remembered that in actual cases these may cause serious errors if neglected.

In most of the problems that follow, (3), (4), and (5) are neglected, and unless otherwise stated, friction is also negligible.

The student should note carefully that all forces which do not cause motion are excluded, as they do no work.

276. Explain why a machine should be of sufficient rigidity that the deformation of its parts should be extremely small.

277. Distinguish between the total energy of a system and its available energy.

278. A railway train, in which the couplings between the cars are heavy springs, begins to move, due to the work done by the engine. State how the energy supplied is being distributed while the train is acquiring speed.

279. If the steam is shut off, from whence comes the energy which keeps the train in motion?

280. What becomes of the potential energy which we store in a watchspring when we wind it?

281. The pitch of a screw is .5 mm. A lever 40 cm. long is used to turn it. A force equal to a weight of 20 kg. applied to the lever will cause the screw to exert what force?

282. Show that the screw is an example of the inclined plane.

283. A lever is 2 m. long, the point of support 30 cm. from the end. A force of 10^8 dynes applied to the long arm will give what force at the short arm?

Consider the work in any displacement. Then

force applied × distance it moves = force exerted × distance moved.

Let the angle turned through $= \theta$.
Distances are $170\,\theta$ and $30\,\theta$.
Work $= 10^8 \cdot 170\,\theta = x \cdot 30\,\theta$.

$$\therefore x = \tfrac{17}{3} \cdot 10^8 \text{ dynes.}$$

284. The radius of the wheel of a copying press is 30 cm. One turn lowers the plate .25 cm. Find the force exerted if the applied force is enough to lift 20 kg.

285. In a hydrostatic press the distances moved by the pistons are in the ratios of 1 to 1000. What is the force ratio?

286. In an ordinary pump handle the long lever arm is 3 ft., the short one 6 in. What force applied to the longer will lift 40 kg. on the shorter?

287. A system of gear wheels is used to raise weights. When the first is turned 360° the last turns 60°. The radius of the first is four times that of the last. What is the force ratio?

288. In the system of pulleys connected as shown in Fig. 29, find the relation between w and W: (a) by principle of work;

(*b*) by considering the tensions of the cords. Neglect the weight of the pulleys.

289. In a system of eight movable pulleys connected as in Fig. 29, find the weight which 20 kg. would lift, neglecting the weight of the pulleys and friction.

290. Find by the principle of work the relation between *w* and *W* when each pulley weighs *p* grams. It is found by experiment that the values of *w* computed above are too small to explain this.

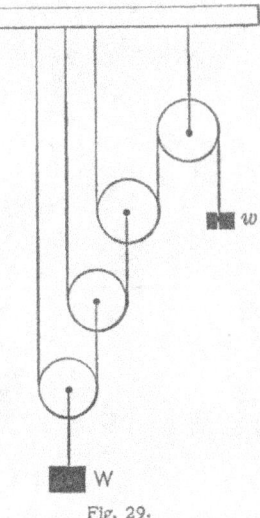

Fig. 29.

291. A system of two movable pulleys, as in Fig. 29, is of negligible friction, and the weight *w* is twice as large as it should be for equilibrium. What will be the acceleration of *w*? of *W*?

292. In a system connected as in Fig. 30, find the relation between *w* and *W*: (*a*) Neglecting weight of pulleys. (*b*) When lower block weighs *M* grams.

293. Find the relation when there are *n* pulleys above and *n* below. When there is one more above than below.

In the wheel and axle we have, if connection is rigid and the cord inextensible, light, and flexible,

Work done by falling of M_1 when angle turned is θ,

$$M_1 R\theta = M_2 \cdot r\theta; \quad i.e. \quad M_1 R = M_2 r,$$

$$\frac{M_1}{r} = \frac{M_2}{R}.$$

(Weights are inversely as radii.)

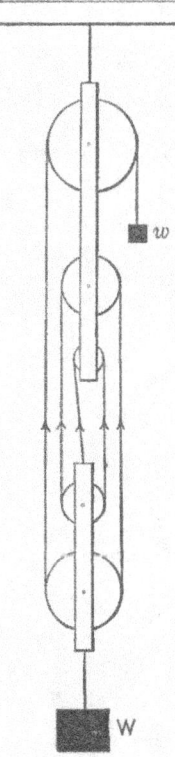

Fig. 30.

For gear wheels we have the same principle. Let R, R_1, and r be the radii of the large wheel, the small wheel, and the axle of the small wheel.

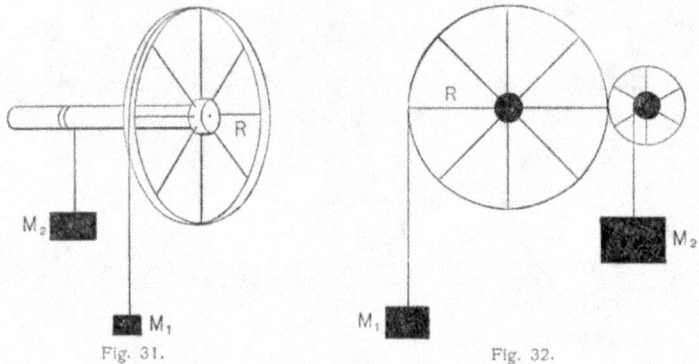

Fig. 31.　　　　　　　　　　　　Fig. 32.

If there is no slipping when R turns through an angle θ, R_1 turns through an angle $\dfrac{R}{R_1}\theta$.

$$\text{Work by } M_1 = M_1 \cdot R\theta.$$

$$\text{Work on } M_2 = \frac{M_2 r \cdot R\theta}{R_1}. \quad \therefore M_1 = \frac{M_2 r}{R_1}.$$

294. If the axle of the wheel (Fig. 33) be 4 cm. in diameter, the mean radius of the wheel 40 cm., the mass of the rim 800 g., the axle and spokes being small in comparison, the mass $M = 200$ g, what will be the velocity of M when it has fallen 4 m.? Forming the energy equation we have, if v is the velocity of M,

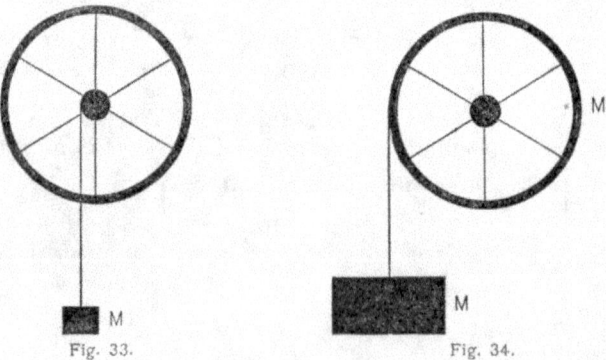

Fig. 33.　　　　　　　　　　　　Fig. 34.

$$E_k = \tfrac{1}{2}\,200 \cdot v^2 + \tfrac{1}{2}\,800\,[20\,v]^2 \quad \text{[Kinetic energy acquired.}$$
$$E_p = 200 \cdot g \cdot 400, \qquad\qquad \text{[Potential energy lost.}$$

Equate and solve for v.

295. A mass M is suspended by a flexible cord wound around a heavy rimmed wheel. The radius of the wheel is R; the mass of the rim M'. What will be the velocity of M after falling a distance h? (Neglecting the spokes.)

Let $v =$ velocity required.

Every particle of the rim is moving with a velocity v.

$$\therefore E_k = \tfrac{1}{2}(M + M')v^2.$$

Lost $E_p = Mgh.$

$$\therefore v^2 = \frac{2Mgh}{M + M'}.$$

296. In Fig. 35,

$M = 8000$ g.

$M' = 200$ g.

$R = 1$ m.

$r = 2$ cm.

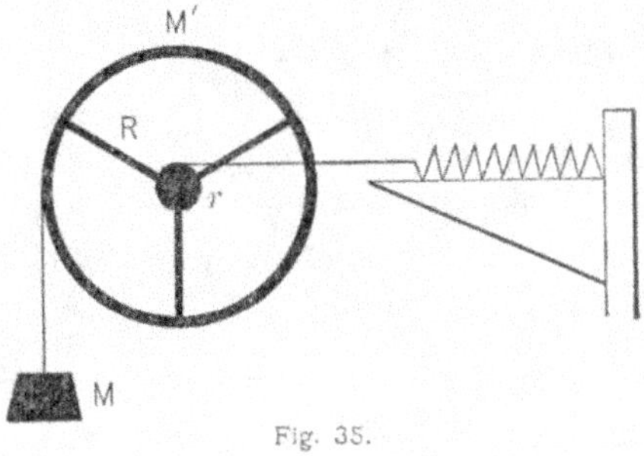

The spring lies on a frictionless shelf, and is connected by flexible thread

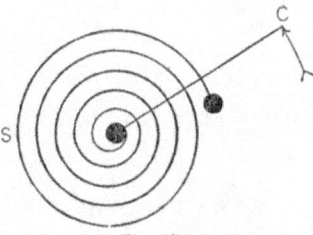

Fig. 35.

to the axle. If M falls 2 m., discuss the energy changes in the system: (1) Neglecting friction of all kinds. (2) When friction is considered constant.

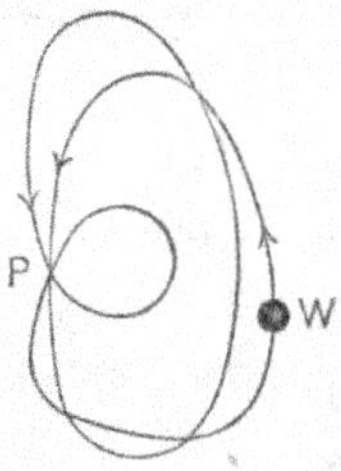

297. A weight W is carried through the point P any number of times. Is its potential energy when at the point P any different at successive times of passage?

Fig. 36.

298. A crank C is turned, thereby "winding up" a spring s. Is the potential energy of the crank dependent only on its position? Explain.

299. A strong rubber band is stretched between two points on a horizontal table A and B. If A remain fixed and B is moved to B'

Fig. 37.

by any path such that the band is straight, show that the work done depends only on $AB' - AB$; *i.e.* on the initial and final positions of the ends.

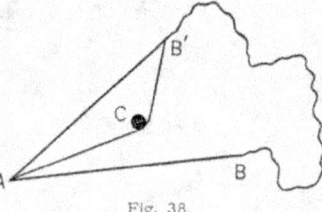

Fig. 38.

300. If the band were drawn around a peg at C, or made to occupy any curved path between A and B, upon what would the work done depend?

301. If the force law is $\dfrac{mm'}{r^2}$, find the work done in carrying m' from r_1 to r_2.

Since the force is not constant, we must divide up the displacement into very short elements, multiply each by the mean force for that element, and add all these results together;

$$i.e.\ dW = \frac{mm'}{r^2}\,dr,$$

or

$$W = \int_{r_1}^{r_2} mm' \frac{dr}{r^2} = mm' \left[\frac{1}{r_1} - \frac{1}{r_2} \right].$$

If $r_2 = \infty$, $^{\infty}W_{r_1} = \dfrac{mm'}{r_1}$, since $\dfrac{1}{r_2} = 0$.

302. How much potential energy will 1 kg. have when it is 1 m. above the sea-level, if we consider its potential energy as 0 when at the sea-level?

303. If g were constant and a surface were drawn everywhere 1 m. from the sea-level, would 1 kg. placed in this surface have a definite potential energy? What would this surface be called? A stone falling freely would strike such a surface at what angle?

304. Explain how the "potential" at a point differs from the potential energy which a mass would have if placed at the point.

305. If two masses attract each other according to the law $\dfrac{mm'}{r^2}$, what will be the force pulling them together when r is infinite?

306. In skating on smooth "level" ice, does one gain potential energy? In climbing an icy hill, is one's potential increased?

307. Are the horizontal floors of a building "equipotential" surfaces?

308. If the work done in carrying 1 kg. from the basement to the first floor is called the potential of that floor, the distance between the floors being uniform, what is the potential of the fourth floor?

309. If the potential of the first floor is 3.10^8, what work will be required to carry 80 kg. from the first to the third floor?

310. In a brick building perfectly built, do the horizontal edges of the bricks lie in equipotential surfaces? Given the potential at the level of one layer, the mass of one brick in a layer, the number of bricks in that layer, how find the work in elevating the whole number?

311. A man walks from a certain point along any path or up hill and down, and returns to his starting-point. What relation exists between the work he has done against gravity, and the work done by gravity on him?

Does it follow that he has done no work? Explain your answer.

312. A man standing on a sloping roof has potential energy. What hinders its transformation into kinetic energy?

313. A body is drawn up a rough inclined plane. Against what forces is work done? State the relation between the energy expended, the potential energy of the body at its highest point, and the work done against friction.

314. How much work is done in taking 80 units of mass from a place where the potential is 5 to one where potential is 1? where the potential is 25?

315. A reservoir on a hill filled with water is said to have what potential? If connected by a pipe with the sea-level, in what direction will water flow?

316. When the potentials at two points very close together are given, how can the force at that point be found?

317. If the potential at points along a certain line is given by $V = f(x)$, find the force function.

318. If $V = f(x)$ between two points x_1 and x_2, and the force is constant, what condition does $\dfrac{d^2 V}{dx^2}$ satisfy between x_1 and x_2?

319. Two cylindrical reservoirs of the same capacity stand on the same horizontal plane; the height of one is four times the height of the other. Which would you prefer to fill with water?

320. When two reservoirs have the same depth of water and one is larger than the other, compare the pressure exerted by each at a given point to which each is connected by a pipe. Compare the potential energy of the two.

321. If the values of a working force are taken as y, and the distance moved as x, what will the area of the surface between two ordinates, the curve, and the axis of x mean?

322. When will the curve be a straight line? What will its slope mean?

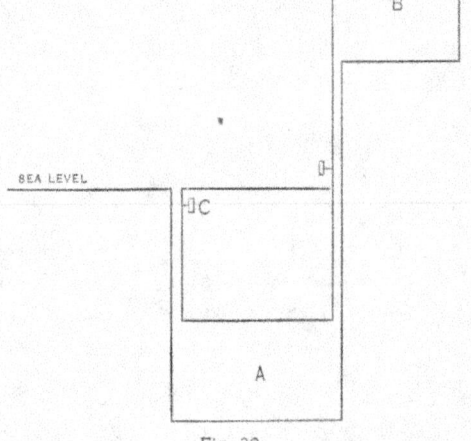

Fig. 39.

323. A constant force acts on a mass subject to friction, the force being greater than the friction. Draw the time-velocity curve (initial velocity o). Discuss the curve, and explain the meaning of its slope, area, etc.

324. A reservoir A is made below the sea-level. What can you say of its potential (taking that of the sea-level as o)? If A and B are connected, is the potential of B altered (c closed)? If A and B are connected and c is opened, what potential changes will occur? (Fig. 39.)

FRICTION

325. Define friction. What do you mean by sliding friction?

326. What are the laws of sliding friction?

327. State what you mean by the coefficient of friction.

328. If a body is "slippery," is the coefficient of friction between it and other bodies large or small?

329. Explain why it is difficult to walk up an icy hill.

330. Explain why rails are "sanded." Why is a violin bow "resined"?

331. A certain force is required to move one surface over another when the pressure between them is P. If P were doubled, what force would be required? if μ were doubled and P were unchanged? if both P and μ were tripled?

332. A mass of 80 kg. on a horizontal plane requires a force equal to the weight of 1.6 kg. to keep it in uniform motion. What is the coefficient of friction?

333. The coefficient of friction between two surfaces is 0.14. A pull of 20 kg. weight will overcome what pressure between the surfaces?

334. If the coefficient of friction is 0.2 between the driving wheel of a locomotive and the rail, what must be the weight, in tons, of the locomotive in order to exert a pull equal to 8.96 T.?

335. The coefficient of friction between a block and a plane is .3. At what angle should the plane be inclined that the block may just slide down it when started? What is the angle named?

336. For a certain plane and block, the coefficient of friction is .2. What force applied parallel to the plane would just draw the block up if it weighs 100 kg., and the plane is inclined 5° with the horizontal?

337. L is a load drawing W along a horizontal plane by means of a cord and pulley, as in Fig. 17.

     If $L = 8$ kg., $W = 40$ kg., pulley friction o; find μ.

     If $\mu = .18$, $W = 80$ kg., pulley friction o; find L.

     If $\mu = .3$, $L = 10$ kg., pulley friction o; find W.

Supposing in each case that the system moves uniformly when started.

338. Solve each part of the preceding example if the co-efficient for the pulley $= .03$.

339. If L were twice as large as specified in 337, find the acceleration.

340. Draw a diagram showing the forces acting when one body is slid uniformly over another.

341. The coefficient of friction between two surfaces is .2. They are inclined at an angle of 60° with the horizontal. What will be the acceleration?

342. A mass of 40 kg. is placed on a plane inclined 50°. The coefficient of friction is .3. What force will be required to draw the mass up the plane with an acceleration of 100 cm. per second per second?

343. If a series of observed values of L and W were used as co-ordinates, what kind of a line would result?

344. If in determining μ by the horizontal plate method the cord passing over the pulley is not parallel to block, show how the correct value of μ may be found.

345. Find the direction and magnitude of the least force required to drag a heavy body up a rough inclined plane. What is the result if the plane is horizontal?

346. A block of weight W rests on a horizontal plane; an elastic spring is used to draw it along at a uniform rate. If the angle at which the elongation of the spring is least is ϕ, find the coefficient of friction.

347. A force of $8 \cdot 10^5$ dynes acts for 1 min. on a mass of 1 kg. sliding on horizontal surface. The velocity acquired was $3 \cdot 10^4$. What was the coefficient of friction?

348. A long plank lies on a nearly smooth inclined plane. A man attempts to walk up the plank. What happens?

G

PENDULUMS. MOMENTS OF INERTIA

349. Find the time of vibration of the following simple pendulums : $[g = 980]$; $l = 16$ cm., 32 cm., 36 cm., 9 cm.

350. A heavy sphere of small radius is suspended by a thread 5 m. long. How many times will it vibrate in an hour ?

351. What must be the ratio of the lengths of two simple pendulums that one may make three vibrations while the other makes four ?

352. A seconds pendulum loses 8 sec. per day when carried to another station. Compare the values of g at the two places.

353. A pendulum is carried upward with an acceleration equal to g. What will be the effect on its period ? What would be the effect if it moved downward with the same acceleration ?

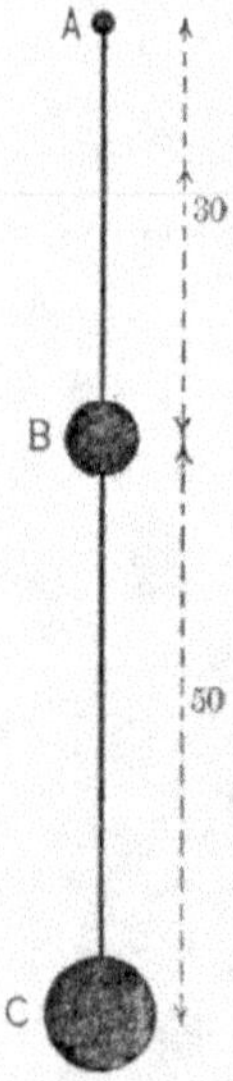

354. AC is a light rigid rod suspended at A. B and C are two small heavy spheres attached to the rod.

$$AB = 30 \text{ cm.}, \quad AC = 80 \text{ cm.}$$

Mass of B, 20 g. ; mass of C, 50 g.

(*a*) Find the periodic time of each if the other were absent.

(*b*) Find the periodic time of the system.

The expression $\tau = 2\pi\sqrt{\dfrac{\Sigma mr^2}{MgR}}$ becomes, in this case,

$$\tau = 2\pi\sqrt{\frac{20 \cdot 30^2 + 50 \cdot 80^2}{(20 + 50)980 \cdot 66}} \text{ (approx.).}$$

Taking the numerator and dividing it by the total mass, we have K for this case.

Hence if R had been given, the actual masses need not be known.

Fig. 40

355. In the system shown in Fig. 21, all lengths are measured from S.

Find (a) the Σmr^2 ;

      (b) the distance from S to center of gravity.

      (c) the periodic time of the system.

(Neglect weight of the rod.)

356. Find the time of vibration of a compound pendulum consisting of a uniform cylindrical rod 2 m. long, radius 2 cm., knife edges 40 cm. from end.

357. If $\tau = \pi \sqrt{\dfrac{\Sigma mr^2}{MgR}}$, what do you mean by R? Name two values of R which could not be used in finding g.

358. Find the moment of inertia of a thin uniform rod :

(a) When the axis is $\perp$ to end of rod.

(b) When the axis is $\perp$ to middle point.

$$\left[\Sigma mr^2 \text{ becomes } \int_0^l \rho x^2 dx. \right]$$

What is the relation between these two values and the center of gravity of the rod?

359. Find the moment of inertia of a thin rod whose density increases uniformly from one end to the other :

(a) When axis is $\perp$ to light end.

(b) When axis is $\perp$ to heavy end.

(Note that $\rho = \rho_0 + kx$.)

360. What relation exists between the two values above and the energy which the rod would have with a given angular velocity in the two cases?

361. Find the moment of inertia of a rectangular area, axis through the center and in the plane of the figure parallel to one side.

362. Find the moment of inertia of a thin circular plate, axis any diameter.

363. Find the moment of inertia of a circular plate of uniform density, axis through center and perpendicular to plane of the circle.

364. Find the moment of inertia of a circular plate, axis perpendicular to plane of circle and through its center, when the density increases uniformly from the center outward.

365. Find the moment of inertia of a right circular cylinder, axis through center and perpendicular to axis of the cylinder, length of cylinder l.

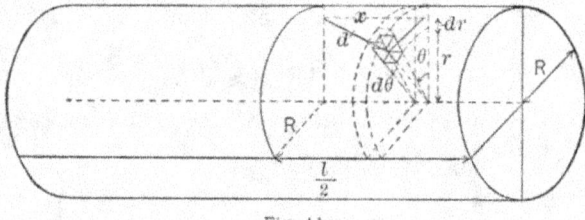

Fig. 41.

By direct integration we may consider the volume element as having a base $r d\theta dr$, and a thickness dx.

Then
$$dm = \rho r\, dr\, d\theta\, dx,$$

$$d^2 = x^2 + r^2 \sin^2\theta,$$

$$\Sigma m d^2 = 2\int_{\theta=0}^{\theta=2\pi}\int_{r=0}^{r=R}\int_{x=0}^{x=\frac{l}{2}}\rho[x^2 + r^2\sin^2\theta]r\,dr\,d\theta\,dx$$

$$= M\left[\frac{l^2}{12} + \frac{R^2}{4}\right].$$

It may be observed that this result is the sum of two parts, the first the same as Ex. 358 (*b*), the second the same as Ex. 362. The energy of the rotating cylinder is, in fact, made up of two parts, one due to the motion of the center of gravity of each circular lamina, the other due to the rotation of these laminæ about their diameter with the *same angular velocity* as the axis of the rod.

In all cases of finding moment of inertia, we have to express $\Sigma m r^2$ as an integral whose form and limits are determined by the problem in hand. It should be remembered by the student in physics that energy of rotation is the thing of real interest and importance rather than the particular mathematical machinery involved.

ELASTICITY

366. Define elasticity of solids; of fluids.

367. When is a body said to be highly elastic and when inelastic? To which of these classes does rubber belong? glass?

368. State what is meant by the term **stress.** What is the stress when 40 kg. rests on a horizontal surface 10 cm. square?

369. A vertical rod 4 sq. cm. cross-section sustains a weight of 100 kg. What is the stress?

How would the stress be changed if the weight were doubled and the cross-section halved?

370. Define and illustrate the term **strain.**

371. A rod 1 m. long is stretched so that its length is 100.04 cm. What is the strain?

372. A cube 20 cm. edge is compressed so that its volume is 7995 c.c. What is the strain?

373. What is meant by the term **elastic limit?**

374. What sort of a curve would represent Hooke's law?

375. A series of weights are suspended by a wrought iron wire. The ratio $\dfrac{\text{increase of length}}{\text{original length}}$ is taken as x, and $\dfrac{\text{force applied}}{\text{area of cross-section}}$ as y. Fig. 42 shows the result. What does the straight portion OB represent? What does the slope

85

of that portion mean? Estimate the safe load. What does
the bend indicate?

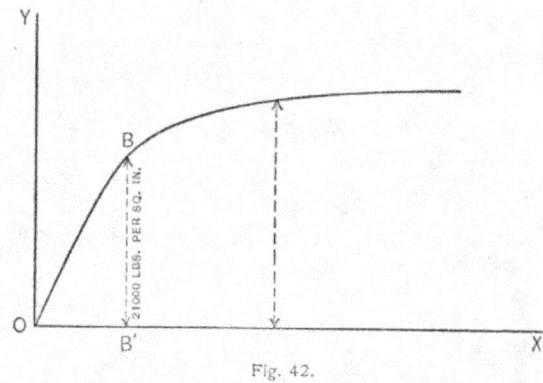

Fig. 42.

376. Define Young's Modulus. It was found that if the
elastic limit would permit so great an extension, it would
require a force of $17 \cdot 10^{11}$ dynes per unit area of cross-section
to double the length of an iron rod. What was Young's
Modulus?

377. Taking Young's Modulus for iron as $2 \cdot 10^{12}$, find the
increase in length of an iron wire 3 m. long when stretched by
a force equal to the weight of 4.5 kg., the radius of the wire
being .5 mm.

378. What effect will stretching a wire have on its radius?

379. A glass tube is stretched in the direction of its length,
would its capacity be changed, and if so in what way?

380. A circular cylinder AB, Fig. 43, is rigidly clamped at
A, and a twist can be given to it by a wheel and weight as
shown. A series of pointers are fastened at points distant
$\dfrac{l}{8}, \dfrac{l}{4}, \dfrac{3l}{8}$, etc., from A.

(*a*) If the wheel is turned 16°, through what angle would each
pointer turn?

(*b*) If *M* was 10 kg. in case (*a*), what would be the twist produced by 25 kg.?

(*c*) If *M* were as in case (*a*) and *R* were multiplied by $2\frac{1}{2}$, how would the distortion compare with that in *b*?

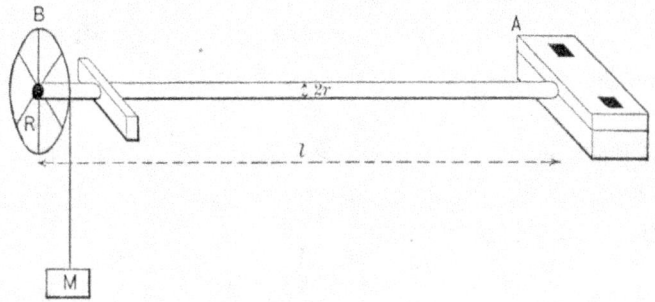

Fig. 43.

(*d*) If the length were half as great, compare the moments required to turn the wheel through the same angle.

(*e*) If the radius of the cylinder were reduced one-half, how would the angles mentioned in *a* be altered if the length and the moment of the applied force were unchanged?

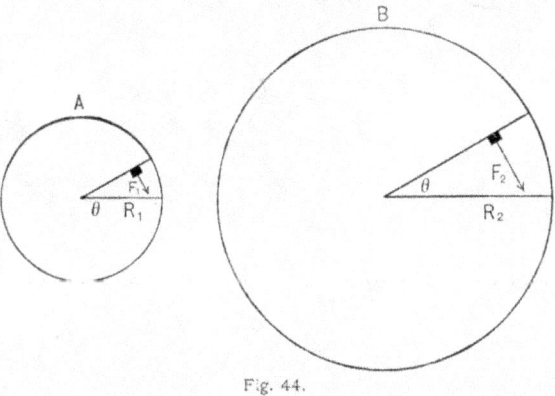

Fig. 44.

381. If *A* and *B*, Fig. 44, are the cross-sections of two circular cylinders of the same material and length, the free end of each is twisted through the same angle θ.

Compare (*a*) the number of elements of area displaced.

(*b*) the mean displacement of these elementary areas.

(*c*) the mean return forces per unit area.

(*d*) the mean leverages for these return forces.

(*e*) the total torques or moments tending to restore the cylinders to their former positions.

382. How does the torque vary with the length of the cylinder?

383. By reference to 380 and 381, find the moment of torsion for a brass wire 3 m. long, 5 mm. radius, given the coefficient of rigidity for brass $= 38 \cdot 10^{10}$.

$$\text{Show that } T = \int_0^R n \frac{2\pi x^3 dx}{l}, \text{ etc.}$$

384. The moment of torsion of a wire 240 cm. long, radius .7 mm. is 17.7. What force applied 2 cm. from its axis and perpendicular to a radius would twist one end of a meter length of this wire 360°?

LIQUIDS AND GASES

385. Distinguish between a liquid and a gas.

386. State fully the reasoning by which the following conclusions are reached :

(*a*) At any point in a liquid at rest the pressure is equal in all directions.

(*b*) The pressure at any point on a submerged surface is normal to that surface.

387. Show that the intensity of pressure in a homogeneous heavy liquid varies directly as the depth.

388. Explain what is meant by a "head" of *h* feet of water, a pressure of *h* cm. of mercury.

389. Express a pressure of 100 lb. per square inch in kilograms per square meter.

390. Is it essential that a barometer tube be of uniform bore ?

391. A barometer tube inclined from the vertical by 5° reads 765 mm.   Find the correct reading.

392. Compute the height of the "homogeneous atmosphere" when the barometer stands at 740 mm.

393. Express in atmospheres the pressure existing at a depth of 20 m. in sea water.

394. Find the pressure at a depth of 6 cm. in mercury surmounted by 4 cm. of water of unit density ; and this, again, by 12 cm. of oil of density .9, atmospheric pressure not considered.

395. Neglecting atmospheric pressure find the intensity of pressure due to a head of 10.37 m. (34 ft.) of water; (*a*) in grams weight, (*b*) in dynes.

396. Find in centimetres of mercury the pressure at a depth of 20 m. in water of unit density, the barometer standing at 76 cm.

397. The pressure at the bottom of a lake is 3 times that at a depth of 2 meters, what is the depth of the lake?

398. At what depth in mercury will be found a pressure equal to that existing in sea water at a depth of 1 km.?

399. The air sustains a column of water 33 ft. (10.0 m.) high. To what internal pressure is the tube of a syphon subjected at a height of 30 ft. above the reservoir?

400. Explain the action of an ordinary suction pump. What is the maximum theoretical height to which water can be raised by such a pump?

401. A body of volume 24 cc. weighs in air at 0° and 760 mm. 16.142 grams. Correct the reading for the weight of displaced air, neglecting the air displacement of the weights.

402. Two liquids that do not mix are contained in a U tube, the difference of level is 4 cm., and the distance between the free surface of the heavier liquid and their common surface is 6 cm. Compare their densities.

403. A U tube 16 cm. high contains mercury to a height of 4 cm.; how many centimeters of chloroform can now be poured into one arm?

404. Alcohol is poured into one arm of a U tube containing mercury; when equilibrium obtains it is found that the free surface of the alcohol is 17 times as high as that of the mercury above the common surface of the two liquids; what is the density of alcohol?

405. Find the pressure on the upper surface of a horizontal plane 12 cm. square when immersed to a depth of 30 cm. in a solution of density .12.

On every square centimeter of the plane the pressure is the weight of a column of the solution 1 sq. cm. in section and 30 cm. high plus the pressure of the atmosphere on 1 sq. cm. of the free surface. This gives as total pressure on one side of the plane, the barometer reading 76 cm.

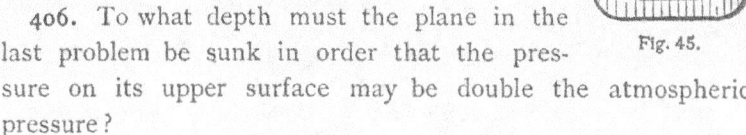

$$144\,[(30 \times 1.2) + (76 \times 13.6)] =$$

The pressure on the under surface of the plane is equal and opposite to this.

Fig. 45.

406. To what depth must the plane in the last problem be sunk in order that the pressure on its upper surface may be double the atmospheric pressure?

407. A square of area 1.24 sq. m. has its upper edge in the free surface of a body of water and its lower edge 80 cm. below the free surface. Find the liquid pressure upon one side of it.

Note that here we have an intensity of pressure varying uniformly from zero at the surface of the liquid to a maximum at the lower edge of the area. We need to find the *mean* intensity of pressure.

408. By what law would the pressure on the area mentioned in the last problem vary with its inclination to the free surface?

409. Sketch the form of a dish such that the hydrostatic pressure on its bottom shall be (*a*) greater than, (*b*) equal to, and (*c*) less than, the weight of the contained liquid.

410. A hole 15 cm. square is punched in the hull of a sea-going vessel at a depth of 3.2 m. below the surface of the water. Compute the force necessary to hold a board over the opening.

411. The water in a pond is confined by a dam of rectangular surface. After heavy rains the water rises by $\frac{1}{2}$ its normal height, although still not overflowing the dam, the surface area of the pond increases at the same time twofold. How does the total pressure on the dam vary?

412. Find the total pressure on a rectangular sluice-gate 8 ft. wide and 6 ft. deep when the water stands at a height of 5 ft.

413. Find the *center of pressure* of a rectangle whose upper edge is in the free surface of the liquid.

The resultant pressure does not pass through the geometrical center of the rectangle because the distribution of pressure is not uniform but varies as the depth. Let b = the breadth of the rectangle. Imagine the total fluid pressure on the right of the rectan-
gle to be concentrated at a certain point distant x from
the surface. Then if we imagine equilibrium to still
exist, we must have the sum of the moments of the
various pressures about the upper edge as an axis $= 0$.
The pressure on a horizontal strip dh wide and b long
is $h.bdh$. Its moment about the upper edge is $hbdh$.

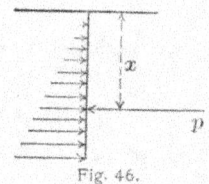

Fig. 46.

Summing these moments, together with the moment of P, which is negative, we have

$$b\int_0^h h^2 dh = Px$$

$$\frac{bh^3}{3} = Px$$

Remembering that $P =$ (mean depth) × area,

we have finally $x = \frac{2}{3}h$.

414. Find the *center of pressure* of a rectangle whose upper edge is horizontal but submerged to a depth of h_1.

415. If the rectangle were inclined at an angle a to the surface of the liquid, would the *center of pressure* change?

416. A right cone, vertex upward is filled with water. Show that the resultant pressure on the curved surface is equal to twice the weight of water in the cone and acts vertically upward.

The volume of the cone is equal to $\frac{1}{3}$ the volume of a right cylinder of the same base and altitude. If such a cylinder be placed over the cone, and the space between it and the conical surface filled with water and the water inside the cone removed, the pressure on the curved surface would remain unaltered. Using this fact the proposition is readily proved.

417. The diameter of the small plunger of a hydrostatic press is 8 cm. That of the large plunger is 1 m. The pressure applied to the small plunger is 260 kg. What load is sustained on the large plunger?

418. The diameters of the two plungers of a hydrostatic press are 4 in. and 3 ft., both being circular. The smaller plunger is worked by a lever whose arms are in the ratio 10 : 1. Find the total load that can be lifted by a man exerting a force of 120 lb.

SPECIFIC GRAVITY AND PRINCIPLE OF ARCHIMEDES

419. A man can just lift a cylindrical jar when filled with water. How many men would be required to lift the same jar filled with a liquid of sp. gr. 12?

420. To what height could the jar be filled with mercury in order that one man could just lift it?

421. Why is it easier to swim in salt than in fresh water?

422. Explain why a balloon filled with hot air rises.

423. Four spheres of the same size are made of Pt, Pb, Ni, and Al respectively. Compare their weights.
If of the same weight, compare their radii; their volumes.

424. A gold and a silver coin are exactly similar in form and of equal weight. What is the ratio of their volumes?

425. Explain why the actual intensity of gravity need not be known in finding specific gravity.

426. If a place could be found where g is 0, could specific gravity still be found, and if so, how?

427. Suppose the space V in a liquid (Fig. 47) to contain matter of steadily increasing density. At first one-tenth that of the liquid, and finally ten times as dense. Show how the resultant force should vary. Draw a curve using density as x, and resultant force on V as y.

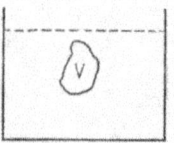

Fig. 47.

428. A bottle filled with water weighs 172 g.; the bottle weighs 72 g. What will it weigh when filled with sulphuric acid? Mercury? Oil of turpentine?

429. A cube of silver and one of gold are of equal size. Compare their weights. If of equal weight, compare their edges.

430. A body in air weighs 40 g. ; immersed in water, it weighs 30 g. Find its specific gravity.

431. A body weighing 80 g. and sp. gr. 4 is immersed in a liquid sp. g. 2. How much weight does the body lose?

432. A body of volume 8 c.c., sp. g. 6, is immersed in liquid of sp. gr. 4. What is its loss of weight?

433. What force would be required to hold a mass of 80 g., sp. gr. 5, under the surface of a liquid of sp. gr. 13.6?

434. A body weighed in water loses 25 g. ; weighed in a liquid of unknown density it loses 50 g. Find density of the liquid.

435. A body in air weighs 50 g. ; its sp. gr. is 8. When weighed in a liquid, it loses 10 g. What is the specific gravity of the liquid?

436. A body immersed in one liquid loses 20 per cent of its weight; when immersed in a second liquid it loses 40 per cent of its weight. Find the ratio of the specific gravities of the liquids.

437. A sinker in water weighs 40 g., a block of wood in air weighs 30 g. ; both in water weigh 20 g. Find specific gravity of the wood. Draw the force system when both are weighed in water.

438. A cork in air weighs 8 g. ; a sinker in water weighs 60 g. ; both in H_2O weigh 28 g. Find the specific gravity of the cork.

439. The specific gravity of a body is 4. What would be its acceleration due to gravity when in water, neglecting friction? What if specific gravity were .4?

440. A body floating in water is placed under the receiver of an air pump and the air is exhausted. Will the depth to which the body sinks be altered? Explain your answer fully.

441. A sinker, volume 80 c.c., sp. gr. 8, is fastened to a piece of wood weighing 35 g. in air; both in water weigh 525 g. What is the specific gravity of the wood?

442. Does specific gravity depend on the units of mass, etc., employed?

443. A cork, sp. gr. .6 and volume 15 c.c., is attached to a brass sinker, sp. gr. 8. What must be the volume of the brass in order that the combination may just sink in water?

444. What must be the edge of a hollow brass cube 1 cm. thick that will just float in water?

445. A sinker of lead, sp. gr. 11.3, is attached to a fish line weighing .005 g. per centimeter and sp. gr. .1. What must be the volume of the lead to pull 10 m. of the line under water?

446. A uniform rod weighted at the bottom is immersed successively in several liquids whose densities increase uniformly. What will be the relation of the volumes immersed?

447. A block of lead in air weighs 330 g. When suspended in water it is found that the water and containing vessel gains 30 g. in weight. What is the specific gravity of lead?

448. Eighty c.c. of lead, sp. gr. 11.3, 20 c.c. of cork, sp. gr. .2, and 10 c.c. iron, sp. gr. 7.8., are fastened together. What would they weigh in water?

449. Compute the specific gravity of glass from the following data:

Weight of bottle . . . . 20 g.
Weight of bottle and H_2O . . . . 100 g.
Weight of powdered glass . . . . 15 g.
Weight of bottle containing glass and filled up with H_2O . 110 g.

450. A specific gravity bottle is counterbalanced; it is then filled with water, and 19.66 g. more are needed to keep it balanced. When filled with alcohol only 15.46 g. are needed. What is the specific gravity of alcohol?

451. A hydrometer weighing 100 g. sinks to a certain mark in water, and requires 20 g. additional to sink it to the same mark in another liquid. What is the specific gravity of the second?

452. The specific gravity of a block of wood is .9. What proportion of its volume will be under water when it floats?

453. A block of wood, sp. gr. .7, is to be loaded with lead, sp. gr. 11.4, so as to float with .9 of its volume immersed. What weight of lead is required if the wood weighs 1 kg.: (1) When the lead is on the top? (2) When the lead is immersed?

454. Show how to compute the specific gravity of a mixture of two or more liquids when the volumes mixed and their specific gravities are known:

(*a*) When new volume is the sum of the volumes of components.

(*b*) When there is a decrease of volume.

455. Two liquids which do not mix and of specific gravities 2 and 5 are placed in a beaker. A body of unknown specific gravity is observed to sink until .3 of its volume was in the lower liquid. What was its specific gravity?

456. Eight parts by volume of a liquid whose sp. gr. is 6 are mixed with five parts of a liquid sp. gr. 3. Find the specific gravity of the mixture when there is no reduction of volume. Find it when the total volume is reduced 5 per cent.

457. What is the difference between hydrometers of *constant* immersion and those of *variable* immersion?

458. Explain how each is used, giving an example.

459. A Nicholson's hydrometer weighs 100 g. and sinks to a certain point in H_2O when 40 g. are added. It sinks to the same point in another liquid when 20 g. are added. Find specific gravity of second liquid.

460. A long test-tube with mercury in the bottom and of uniform cross-section is used to determine the specific gravity

H

of a number of liquids lighter than water. Show how to *calibrate* when the point to which it sinks in two liquids of known specific gravity is given.

461. A piece of lead, volume 20 c.c., sp. gr. 11.4, is suspended from one arm of a balance and is immersed in oil, sp. gr. .9. From the other end an irregular mass of gold, sp. gr. 19.3, is suspended in turpentine, sp. g. .8. What is the volume of the gold if the beam remains horizontal?

462. A brick, sp. gr. 2, is dropped into a vessel containing mercury and water. Find its position of equilibrium.

463. Two equal cubes of oak and pine respectively are placed in water. The edge of each is 20 cm. What height of each will be above the surface?

464. A cylindrical rod of wood and iron is to be made so as to just sink in water. Specific gravity of wood, .5; of iron, 7.5. The length of the iron rod is 75 cm. How long must the wood be?

465. According to Boyle's law $pv = k$ at constant temperature. Give two definitions of k from a consideration of the formula. Also show graphically the meaning of k.

466. A cylinder 24 in. long contains 2 cu. ft. of air at a pressure of 15 lb. per square inch. The cylinder is slowly pushed in. (*a*) Find the pressure at several points of the stroke and lay them off as ordinates, thus forming a pressure-volume curve with axis as shown. Discuss this curve. (*b*) What is the total pressure on the inner surface of the piston?

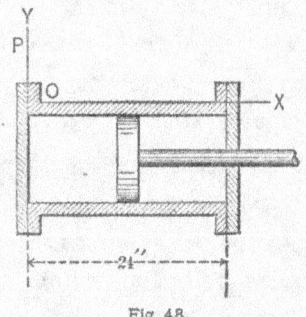

Fig. 48.

467. Show that it follows from Boyle's law that the pressure of a gas at constant temperature must be proportional to its density.

468. Forty c.c. of air are enclosed in an inverted tube over mercury. The difference of level is 50 cm. The tube is depressed until the difference of level becomes 30 cm. What is the volume of the enclosed air?

469. A glass tube 60 cm. long and closed at one end is sunk, open end down, to the bottom of the ocean; when drawn up it is found that the water has wet the inside of the tube to a point 5 cm. below the top; what is the depth of the ocean?

470. An air bubble at the bottom of a pond 6 m. deep has a volume of 0.1 c.c. Find its volume just as it reaches the surface, the barometer showing 760 mm.

HEAT

TEMPERATURE

471. Define *temperature*. Is the sense of touch a reliable measure of temperature ? Explain fully.

472. Bodies at different temperatures are sometimes said to be at different *thermal levels*. What is meant ? Explain the difference between temperature and quantity of heat.

473. What does a mercury-in-glass thermometer really indicate ? How is such a thermometer graduated ?

474. How would you construct a thermometer to be "sensitive"? to be "delicate"?

475. What special advantages does mercury possess as a thermometric substance?

476. If the coefficient of cubical expansion of the liquid in a thermometer is less than that of the envelope, what effect will be produced on heating the thermometer?

477. Reduce to Fahrenheit readings, the following Centigrade temperatures : $45°$, $12°$, $-20°$.

478. Reduce to Centigrade readings the following Fahrenheit temperatures : $212°$, $72°$, $32°$, $-30°$.

479. Plot Centigrade temperatures as abscissas and corresponding Fahrenheit readings as ordinates, and discuss the locus. Also, take from the cross-section paper convenient values, and construct a double thermometer scale; *i.e.* one which gives the temperatures in both systems.

480. At what temperature will both Fahrenheit and Centigrade thermometers give the same reading? What happens to mercury at this temperature?

481. The temperature of a given liquid is taken by both Fahrenheit and Centigrade thermometers. The Fahrenheit reading is found to be double the Centigrade reading. What is the temperature of the liquid in degrees Centigrade?

482. Define the coefficient of linear expansion and establish the formula

$$l_t = l_0(1 + \lambda t),$$

where l_t is the length of a bar of given material at temperature t, l_0 its length at zero, and λ the mean coefficient of expansion for the material between $0°$ and $t°$.

If a bar of given material be heated, it lengthens. Every unit of the original length elongates for every degree rise of temperature an amount λ. This is the *coefficient of linear expansion*. Between narrow limits of temperature the elongation may be taken as proportional to the temperature rise. The *total* elongation for a temperature rise of t degrees from zero must therefore be $l_0\lambda t$, which makes the new length

$$l_t = l_0 + l_0\lambda t = l_0(1 + \lambda t).$$

When t is large, l_t can no longer be taken as a linear function of the temperature, but is represented by

$$l_t = l_0(1 + \lambda t + \lambda' t^2 + \cdots).$$

483. Show that the true linear expansion coefficient at temperature t is given by

$$\lambda = \frac{1}{l_t} \frac{dl}{dt}$$

484. A platinum wire is 4 m. long at $0°$; find its length at $100°$.

We have

$$l_{100} = l_0(1 + .000009\, t)$$
$$= 4 \times 1.0009$$
$$= 4.0036 \text{ m.}$$

485. Show that the value of λ is independent of the unit of length used, but depends upon the thermometric scale used.

486. A lead pipe has a length of 12.623 m. at 15°; find its length at 0°.

487. Why is platinum wire well adapted for use in the "leading in" wires of a glow lamp, or in any circumstances in which it needs to be fused into glass?

488. A certain induction coil has 20,000 turns of copper wire in its secondary coil. If climatic changes cause a rise of 40° in its temperature, express the resulting expansion in turns of mean length.

489. The length of a brass wire at 3° is 12 m.; find its length at 33°.

In this example we might first find the length of the wire at zero degrees, and then by resubstitution find the length at 33°. A sufficiently accurate result, however, is obtained by an approximation. We have

$$l_0 = \frac{l_t}{1 + \lambda t},$$

whence the length at any other temperature t' is

$$l t' = l_t \cdot \frac{1 + \lambda t'}{1 + \lambda t},$$

$$= l_t[1 + \lambda(t' - t)],$$

very approximately when λ is small. [See V.]

490. Assuming that 43° is the maximum temperature to which steel rails, 10 m. long at 0°, are ever subjected during the changing seasons, compute the space which should be left between them when laid at 15°.

491. Measurements are made at 25° upon a brass tube by a steel meter scale, correct at 0°. The result is 6.426 m. Find the length of the tube at 0°.

One should here consider that the result of these measurements is a *number* which shows the ratio of the length of the tube to the length of the scale at the temperature at which the measurements are made. Since the length of the tube at zero is required, the number obtained is too large because of the expansion of the thing measured and too small because of the expansion of the unit. The result sought will therefore be found by multiplying the number by the ratio of the expansion factor of steel to the expansion factor of brass.

492. A brass rod is found to measure 100.019 cm. at 10° and 100.19 cm. at 100°. Find the mean coefficient of linear expansion of brass between 10° and 100°.

The student should work this example first by the accurate method and then by use of the approximate formula (see V.) and compare the results.

493. A platinum bar originally at 15° is placed in a glass-blower's furnace. The increase in length is .96 per cent. Find the temperature of the furnace.

494. When it is desired that a point p shall remain at a constant distance d from a support, an arrangement built on the principle shown in the figure may be used. The rods a, a, and b are of one metal and the rods c, c, are of another. This principle is used in the "gridiron" clock pendulum. Derive the conditions for compensation.

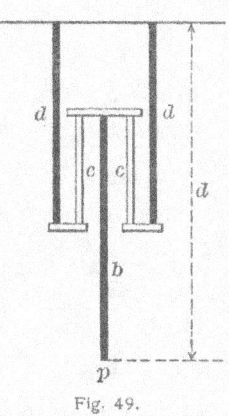

Fig. 49.

495. A lever at A controls a distant railway signal at B. If A and B are connected by a rod, changes in temperature may cause a movement of the signal independent of any motion of the lever. Devise a scheme by which this may be avoided, the same rod being retained.

496. A clock which keeps correct time at 22° has a pendulum made of iron. If the temperature fall to − 8°, how many seconds per day will the clock gain?

NOTE. — The time of vibration of a pendulum is proportional to the square root of its length.

497. Show that if λ be taken as the coefficient of linear expansion of a given material, the coefficient of volume expansion of the same material is approximately 3λ. [See V.]

498. A silver dish has a capacity of 1.026 l. at 75°; at what temperature will its capacity be just one liter?

499. A steel boiler has a surface area of 9.2 sq. m. at 6°; find the per cent increase in this area for a rise in temperature of 80°.

500. Find the mean coefficient of volume expansion of tin on the Fahrenheit scale.

501. Explain how density varies with temperature, and show that when t is small

$$\delta_t = \delta_o (1 - \beta t);$$

and further that

$$\frac{\delta_t}{\delta_{t'}} = [1 + \beta(t' - t)].$$

NOTE. — These results are obtained by approximate methods.   [See V.]

502. The density at 0° of a specimen of wrought iron is 7.3, and the density at 0° of a specimen of tin is 7.4; at what temperature will these two specimens have the same density?

503. Distinguish between *real* and *apparent* expansion of liquids. Show that the coefficient of real expansion of a liquid is sensibly equal to the coefficient of apparent expansion together with the coefficient of cubical expansion of the envelope.

504. The coefficient of apparent expansion of mercury in glass is $\frac{1}{6500}$; the coefficient of real expansion of mercury is $\frac{1}{5500}$. Find the coefficient of volume expansion of glass.

505. A graduated glass tube contains 40 c.c. of mercury at 0°. If the whole be heated to 32°, what is the apparent volume of the mercury?

If glass and mercury had the same coefficient of expansion, the apparent volume would remain unaltered. But taking the expansion coefficient of mercury at 182×10^{-6} and that of glass at $3 \times 85 \times 10^{-7}$, it is evident that the volume of the mercury increases more rapidly than the volume of the tube. This means that the apparent volume of the mercury will increase.

506. A glass flask holds 842 g. of mercury at 0°. How much will overflow if the whole be heated to 100°?

507. Taking the density of mercury at 0° at 13.6, calculate the density at 200°.

508. Taking the density of mercury at 60° as 13.45, find the density at 100°.

509. It is desired to study the true expansion of water. If the proper amount of mercury be placed in a glass bulb, the expansion of the mercury, for any rise of temperature, will equal that of the bulb itself. The volume above the mercury will thus remain constant, and may be filled with water. Any observed increase in the volume of water must therefore be its true expansion. What fraction of the volume of the bulb at zero must be filled with mercury to secure this result?

510. Describe the manner in which water behaves between zero and 10°.

511. The surface of a pond of water is observed to be just freezing. Would you expect the water at the bottom of the pond to be at the same temperature and density as that at the top?

512. Describe the *weight thermometer*. The bulb of a thermometer contains 2.4 kg. of mercury at 0°. The whole is heated to $t°$, causing an overflow of 40 g. Required t.

Let M = total mass of mercury.

m = overflow.

δ = density of mercury.

κ = coefficient of expansion of glass.

a = coefficient of expansion of mercury.

Now the volume of the thermometer at 0° is

$$\frac{M}{\delta},$$

which becomes, at $t°$,

$$\frac{M}{\delta}(1 + \kappa t).$$

The mass of mercury filling the thermometer at $t°$ is

$$M - m,$$

its volume at 0° is

$$\frac{M - m}{\delta},$$

and this volume expands at $t°$ to

$$\frac{M-m}{\delta}(1 + at).$$

But the volume of the expanded mercury is the same as that of the expanded bulb, from which relation t is readily found.

513. A weight thermometer containing 1 kg. of mercury at 0° is placed in an oil bath, and the mass of expelled mercury is found to be 26.4 g. Find the temperature of the bath, the coefficient of apparent expansion of mercury in glass being $\frac{1}{6500}$.

514. What is the law of the expansion of the permanent gases with rise of temperature? Through what range of temperature must a mass of gas be heated, at constant pressure, in order to double its volume?

515. If Charles' law be assumed to hold true for all temperatures, what happens at $-273°$? What is this temperature called? If temperatures be reckoned from this point, how is the expression for the law modified?

516. A mass of gas at 15° occupies 120 c.c. Find its volume at 87°, the pressure remaining constant.

We have according to Charles' law,

$$\frac{V}{V'} = \frac{273 + t}{273 + t'}.$$

$$V = 120 \times \frac{360}{288} = 120 \times 1.25$$

$$= 150 \text{ c.c.}$$

517. Take volumes as ordinates and temperatures as abscissas, and give a graphical representation of Charles' law.

518. At what temperature will the volume of a given mass of gas be three times what it is at 17°?

519. A volume of hydrogen at 11° measures 4 l. If the temperature be raised, at constant pressure to 82°, what is the change in volume?

520. The temperature of a constant volume of gas is raised from 0° to 91°. Find the per cent increase in pressure.

521. Show that for a given mass of gas the quantity $\frac{pv}{T}$, or $\frac{\text{pressure} \times \text{volume}}{\text{absolute temperature}}$, is invariable.

522. Find the dimensions of the product pv.

523. Find the volume of 2 lb. oxygen at a pressure of 3 atmospheres and temperature 27°, the volume of 1 lb. oxygen at 0° and 1 atmosphere being 11.204 cu. ft.

The volume at 0° and 1 atmosphere is

$$v_0 = 2 \times 11.204 \text{ cu. ft.}$$

If the gas is heated at constant pressure to 27°, it expands by Charles' law to

$$v' = \tfrac{300}{273} \times 2 \times 11.204 \text{ cu. ft.}$$

Now if the pressure be increased three-fold at constant temperature,

$$v'' = \tfrac{1}{3} \times \tfrac{300}{273} \times 2 \times 11.204 \text{ cu. ft.}$$
$$= 8.2 \text{ cu. ft.}$$

524. Find the numerical value of $\frac{pv}{T}$ for a mass of 1 g. of air.

Now $\frac{pv}{T} = \frac{p_0 v_0}{T_0}$, where v_0 is the volume of 1 g. at 0° and p_0 is a pressure of 1 atmosphere.

$$p_0 = 13.596 \times 76$$

in grams' weight per square centimeter

$$T_0 = 273°.$$
$$v_0 = \frac{1}{.001293} \text{ c.c.}$$

Therefore, $$\frac{p_0 v_0}{T_0} = \frac{13.596 \times 76}{273 \times .001293} = 2927.$$

525. Compute the value of $\frac{pv}{T}$ for a gas s times heavier than air, of which m grams are taken. Show that the value of this constant depends on the quality and quantity of the gas used.

526. The pressure on a given mass of gas is doubled, and at the same time the temperature is raised from 0° to 91°. How is the volume affected?

527. The pressure of a given mass of air is that due to 120 cm. of mercury, its volume is 1000 cu. cm., and temperature 15°. If now the pressure be increased to 250 cm., the volume becomes 300 c.c. ; what is the temperature ?

528. Find the value of $\dfrac{pv}{T}$, where p is measured in pounds per square foot, v in cubic feet, and T in Fahrenheit degrees.

529. For a certain mass of air $\dfrac{pv}{T} = 58540$. Find its volume at 0° and 760 mm.

530. Show that the final temperature resulting from mixing M grams of a substance of specific heat c and at a temperature T with m grams of water at a temperature t is

$$\theta = \frac{McT + mt}{Mc + m}.$$

531. Solve the equation of 530 for the specific heat c, and extend the problem to the case in which the thermal capacity of the calorimeter is considered.

SUGGESTION. — Some of the heat liberated by the hot body goes to warm the calorimeter, which is assumed to be carried through the same temperature range as the water. This amount of heat is therefore $M_c c' (\theta - t)$, where M_c is the mass of the calorimeter, and c' the specific heat of the material of which it is made.

532. How many minor calories are required to raise the temperature 3 kg. of copper from 16° to 110° ?

533. Equal masses of iron and aluminum cool through the same range of temperature; compare the amounts of heat lost.

534. Assuming no loss of heat, how much heat must be imparted to 2 gal. of water, initially at 14°, in order to raise it to the boiling-point?

535. Compare the thermal capacities of equal volumes of gold and aluminum.

536. Three liters of water at 40° are mixed with two at 9°; what is the temperature of the mixture?

537. If one has available water at the boiling-point and water at 5°, what amounts must he take of each in order to form a mixture of 55 l. at a temperature of 20° ?

538. Into 12 kg. of water at 30° are dropped, at the same instant, 1 kg. of copper at 100° and 1.2 kg. of zinc at 60° ; find the resultant temperature.

539. If a calorimeter be made of material of specific heat c', and if it have a mass m', the product $m'c'$ is sometimes called the water equivalent of the calorimeter. What justifies the use of the term ?

540. A copper calorimeter weighs 62 g.; what is its water equivalent ?

541. In determining the water equivalent of a calorimeter the following data are observed :

Weight of calorimeter	52.66 g.
Weight of calorimeter + cold water	302.71
Initial temperature	11°
Temperature of hot water	80°
Final temperature	14.8°
Total weight after addition of hot water . .	317.61

Compute the water equivalent.

542. Compare the result obtained in the last problem with the computed value, assuming the calorimeter to be made entirely of copper.

543. A silver dish weighing 50 g. contains 500 g. of water at 16°; a piece of silver weighing 65 g. is heated to 100° and then plunged into the water; the resulting temperature is 16.50°; what is the specific heat of silver?

544. A mass of 200 g. of copper is heated to 100° and placed in 100 g. of alcohol at 8° contained in a copper calorimeter, whose mass is 25 g., and the temperature rises to 28.5°. Find the specific heat of alcohol.

545. An iron ball is heated to 100° and then dropped in 3 l. of water at 6°, causing a rise of temperature of 2°; what is the diameter of the ball?

545a. The specific heat of most substances is not a constant, but is a function of the temperature. If the quantity of heat necessary to raise one gram of a substance from 0° to t° be given by

$$Q_t = at + bt^2 + ct^3,$$

show that the specific heat at a temperature t° is

$$C = a + 2bt + 3ct^2,$$

and that the mean specific heat between t° and t'° is

$$C_m = a + b(t + t') + c(t^2 + tt' + t'^2).$$

546. One starts with 100 g. of water at 10°, and to this one adds successive amounts of water from a reservoir maintained always at 100°. Express the temperature of the mixture as a function of the amount of hot water added. Plot a curve between amounts of water added (abscissas) and final temperatures (ordinates). Note the limit beyond which the curve has no physical meaning.

547. Show from the equation for the final temperature in the method of mixtures, that loci similar to that in the last problem are hyperbolas. Discuss fully.

548. Define heat of fusion. What seemed to justify the term *latent heat?*

549. Taking temperatures as ordinates and quantities of heat as abscissas, plot the relation between these quantities for the case in which ice at − 10° is converted into water at 90°.

550. How many calories must be supplied to 15 kg. of ice at 0° to completely melt it?

551. How many grams of ice at 0° must be added to 1000 g. of water at 30° to produce a final temperature of 5°?

552. In a determination of the heat of fusion of ice, the following data are observed :

Weight of calorimeter .	71.5 g.
Water equivalent .	8.5 g.
Weight of calorimeter and water	156 g.
Temperature of water .	54°
Temperature after ice is melted	32°
Weight after addition of ice .	174.5 g.

Compute the heat of fusion of ice.

553. Required, the amount of heat necessary to raise 3 kg. of lead at 10° to the melting-point, and then to melt it.

554. How many grams of lead could be melted by the heat set free, when 160 g. of molten tin solidifies? Each substance is supposed to be at its melting-point.

555. How much ice must be thrown into 6 kg. of water at 41° to produce a final temperature of 8°?

556. Find the least quantity of water at 0° which, surrounding a kilogram of solid mercury at its melting-point (− 40°), will just melt the mercury without altering the temperature of either substance.

557. Find the ultimate common temperature of the ice and mercury in the last problem.

558. What will be the result of mixing 12 kg. of snow at 0° with the same mass of water at 20°? What must the temperature of the water be in order that the snow may entirely melt, the mixture having a temperature of 0°?

559. Show how the specific heat of a solid may be obtained by the use of the ice calorimeter.

560. In a determination of the specific heat of iron a mass of 160 g. is heated to 100° and dropped in the calorimeter. The mass of ice melted is 22.4 g. Compute the specific heat of the sample.

561. A mass of 400 g. of copper is heated in an oil bath and then placed in an ice calorimeter. The mass of ice melted is 150 g. Required the temperature of the bath.

562. It is desired to determine the specific heats of several metals by the ice calorimeter. The samples chosen are of the same mass and are heated to the same temperature, in a bath of boiling water. What mass must be used in order that the computation will be simplified to

$$c = \frac{\text{mass of ice melted}}{100} ?$$

563. Explain the action of freezing mixtures.

564. What is meant by regelation? In what substances should we look for the phenomenon?

565. Explain the making of snowballs, the formation of ice on pavements, and the flow of glaciers, as phenomena of regelation.

566. Why is iron an excellent metal for casting? Why are coins stamped instead of being cast?

567. Punched rifle bullets pursue a straighter course than do cast bullets. What reason can be given for this?

568. What property of wrought iron enables it to be readily welded? How does sealing-wax behave when heated?

569. What amount of heat must be supplied to 10 kg. of water at 100° to convert it into steam at the same temperature?

To convert 1 g. of water at 100° into steam at the same temperature requires 536 calories (heat of vaporization of water). In this case we must have

$$H = 536 \times 10^4 = 5360 \text{ calories.}$$

570. Find the numerical value of the heat of vaporization of water in terms (a) of pound and degree Centigrade units, (b) in terms of pound and degree Fahrenheit units.

571. Explain why evaporation cools. If a few drops of ether be placed on the bulb of a thermometer, an immediate lowering of the mercury is observed; but when the thermometer is dipped in a bottle of the ether, no lowering is observed. Explain.

572. A kettle contains 2 kg. of water at 40°. How much heat must be supplied in order to boil the water away?

573. A calorimeter contains 316 g. of water at 40°. Steam at 100° is passed into the water until the mass of water becomes 336 g. What is the temperature?

The mass of steam condensed is

$$336 - 316 = 20 \text{ g.},$$

which yields the heat of vaporization,

$$20 \times 536 \text{ calories.}$$

Further, the 20 g. of condensed steam in cooling to the final temperature θ yields

$$20 \ (100 - \theta) \text{ calories.}$$

The 316 g. of water originally in the calorimeter is raised from 40° to θ, which means a gain of heat of

$$316 \ (\theta - 40) \text{ calories.}$$

Now equating the heat evolved in condensing and cooling to the heat absorbed by the cool water, the unknown temperature θ is readily found.

574. In a determination of the heat of vaporization of water by passing steam into a calorimeter containing cold water, the following data are obtained:

Weight of calorimeter	71.5 g.
Water equivalent of calorimeter	8.5 g.
Weight of calorimeter and water . . .	173 g.
Temperature of cold water	17°

After passage of steam:

Weight of calorimeter and water . . .	181 g.
Temperature	41°

Compute the heat of vaporization.

575. What is meant by the *total heat of steam?*

576. What amount of steam at 100° must be passed into 16 kg. of water at 0° in which 4 kg. of ice are floating, in order to raise the whole to 30°?

577. Calculate the heat necessary to raise to the boiling-point, and to completely vaporize 120 g. of alcohol at 12°.

I

578. What is meant by a saturated vapor? Upon what does the pressure of a saturated vapor depend?

579. Some values from Regnault's determination of the maximum pressure of water vapor are given below. Plot them.

Temperature (abscissas).		Pressure (ordinates).
0°	. . . .	0.46 cm.
10°	. . . .	0.91 cm.
20°	. . . .	1.74 cm.
30°	. . . .	3.15 cm.
40°	. . . .	5.49 cm.
50°	. . . .	9.20 cm.
60°	. . . .	14.90 cm.
70°	. . . .	23.30 cm.
80°	. . . .	35.50 cm.
90°	. . . .	52.50 cm.
100°	. . . .	76.00 cm.

580. Into a barometer tube in which the mercury stands at 760 mm. a few drops of water are introduced. (*a*) Explain what happens. (*b*) If the temperature be 30°, and there still remain a little water on top of the mercury, what will be the reading of the barometer? (The height of the layer of water is neglected.) (*c*) What are the effects of raising and of lowering the barometer tube, supposing the cistern to be deep enough to admit of this?

581. In a closed chamber saturated water vapor in contact with its liquid exists at a pressure of 23.3 cm. What is the temperature? If means are provided for pumping out the vapor, what happens?

582. How define the boiling-point of a liquid in terms of the pressure of its saturated vapor, and the pressure upon its free surface?

583. How do the results compare with the rise of pressure at constant volume of a gas such as air with increasing temperature? What conclusion can be drawn as to the relative danger from explosion of steam and air engines working at the same temperature?

584. What is the maximum pressure of water vapor at 55°?

585. At Quito, Ecuador, the mean barometer reading is 52.5 cm. What is the boiling-point? How can cooking operations requiring a temperature of 100° be carried on at this altitude?

586. Explain the action of (a) *vacuum pans* for converting sap into sugar; (b) of *digesters* for boiling substances at high temperatures.

587. In a closed vessel is contained water which has cooled so that ebullition has ceased. How may the water be made to boil again without applying heat to the vessel?

588. Give examples of the transference of heat by conduction. Name several metals in order of their conducting powers. What of the conductivity of liquids?

589. A thermometer placed in contact with the different bodies in a cold room shows no variation in temperature, yet some of the bodies feel colder than others. Explain.

590. Why are woolen blankets equally good for keeping the person warm in winter and for preserving ice in summer?

591. Define the coefficient of thermal conductivity.

592. One side of a wall of indefinite extent is maintained constantly at 0°, while the other side is maintained constantly at $t°$. Give reasoning to show that after a certain lapse of time (a) the flow of heat across a section of the wall parallel to the faces is the same as that across any similar section; and (b) that the rate of fall of temperature across the wall is uniform.

593. Show that the dimensions of k, thermal conductivity, are, in thermal units, $ML^{-1}T^{-1}$. Whence, given that the conductivity of silver in C.G.S. is 1.3, find the corresponding value in terms of the pound, foot, and minute. Explain how it happens that *k* thus measured is independent of the thermometric unit.

594. What would be the thickness of a plate of iron that would permit the same flow of heat as a plate of glass 0.3 cm. thick, the areas and temperature difference between faces being the same?

595. What would be the disadvantages of a thermometer whose bulb contained a very large amount of mercury?

596. A coil of copper wire lowered over the flame of an alcohol lamp will extinguish it. Explain.

597. What is the function of the wire gauze in a miner's safety lamp?

598. If 1,440,000 calories pass in 1 hr. through an iron plate 2 cm. thick and 500 sq. cm. in area, when the sides are kept at 0° and 10°, compute the thermal conductivity of iron.

599. The surface of a pond is coated with ice 18 cm. thick. The temperature of the air is − 12°. Compute the amount of heat passing upward through a surface of 1 sq. m. in 1 hr.

Be careful to use consistent units. If .003 be taken as the thermal conductivity of ice, C.G.S. units must be used throughout.

600. The last problem is to be worked on the assumption that the thickness of the ice does not increase sufficiently in one hour to appreciably change the flow of heat. As a matter of fact the ice is growing thicker at a rate proportional to the extraction of heat from the water. Find the law by which the thickness of ice increases with time, temperature remaining as above stated.

601. What is meant by the transfer of heat by convection? Which plays the greater part in the heating of a room, convection or conduction?

602. Explain the method of heating buildings by hot water.

603. Give examples of the modification of climate by ocean convection currents.

604. What is meant by *radiation?* Draw a curve showing the distribution of energy in the visible and non-visible spectra.

605. What class of bodies are good reflectors of radiant heat? good absorbers?

606. Explain how the specific heat of a substance may be determined by the *method of cooling.*

607. Equal masses of water and alcohol cool successively through the same range of temperature in the same dish in times whose ratio is $\frac{100}{45}$. Compute the specific heat of alcohol for the range of temperature used in the experiment.

608. What is meant by the *radiation constant* of a calorimeter? How is it determined experimentally, and how is it used in a specific heat determination by the method of mixtures.

609. What is meant by the term *mechanical equivalent of heat?* Describe any method by which it has been determined.

610. Express 20 calories in ergs.

From Introduction, we take as the value of J, 4.2×10^7 ergs. Hence

$$20 \text{ calories} = 8.4 \times 10^8 \text{ ergs.}$$

611. Show that the numerical value of J in gravitational units varies as $\dfrac{\text{unit of temperature}}{\text{unit of length}}$.

612. To raise 1 gr. of water 1° C. requires 4.2×10^7 ergs. Find the number of foot-pounds required to raise 1 lb. of water 1° F.

613. In a certain machine the power wasted in friction is 21 kilogram-meters per hour. How much water per hour could be heated from 0° to 100° by this amount of power?

614. With what speed should ice at 0° be fired against an impenetrable wall in order to be completely melted, assuming that no heat is lost?

615. Why does the specific heat of a gas at constant pressure differ from the specific heat at constant volume?

616. Describe an experiment to show that air is not cooled by expansion if no external work is done. Is this result true of all gases?

617. A cubic meter of air at 0° and 76 cm. pressure is contained in a cylinder whose piston moves without friction. If the air be heated to 100°, what is the external work done?

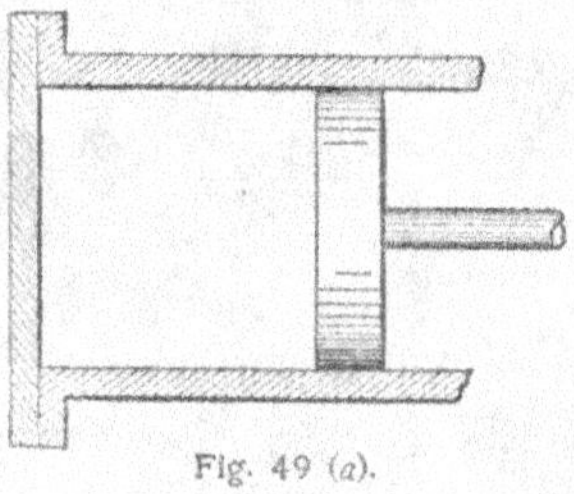

By the conditions of the problem, external work is done against the pressure of the atmosphere. This pressure is

Fig. 49 (a).

$$p = 76 \times 13.6 \text{ grams' weight per square centimeter.}$$

Since the gas expands at constant pressure, the increase in volume is

$$v = \frac{100}{273} \times 10^6 \text{ c.c.}$$

Whence the work is

$$pv = 76 \times 13.6 \times \frac{100}{273} \times 10^6 \text{ gram-centimeters.}$$

618. Compute the heat supplied to cause this expansion.

This is readily done by finding the mass of the air in the cylinder and using the specific heat at constant pressure.

619. Compute the heat required to raise the temperature of this mass of air at constant volume.

620. One liter of air at 0° is confined by a weightless piston in a cylinder whose sectional area is 1 sq. dm. The pressure of the atmosphere is 76 cm. The temperature of the gas is raised to 273°, thus increasing the volume to 2 l. Compute the mechanical equivalent of heat. [Ratio of specific heat at constant pressure to specific heat at constant volume = 1.41.]

621. What is an isothermal line? an adiabatic line? Why is the adiabatic line through any point of the pressure-volume diagram steeper than the corresponding isothermal?

622. Sketch an indicator diagram made up of two isothermals crossed by two adiabatics. Discuss the four steps which are made in carrying the working substance through this cycle.

623. Find the work done on the piston of a steam engine after cut-off, *i.e.* after the entrance port of the cylinder is closed, when the expansion is assumed to take place in accordance with Boyle's law, the back pressure being zero.

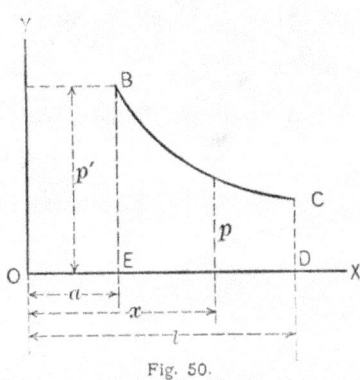

Fig. 50.

Let the positions of the piston at different times be laid off along OX and the corresponding pressures along OY. At E, when the piston has proceeded a distance a, cut-off occurs, after which the pressure falls along BC. Our problem is to find the work corresponding to the area $BCDE$.

If the area of the piston is A, the pressure upon it when it has proceeded a distance x is pA. If it move under this pressure, a small distance dx, the work done is

$$dw = pA\,dx,$$

and the total work corresponding to a distance $l - a$ is

$$W = A\int_a^l p\,dx.$$

But the condition that $pv = $ constant gives

$$p'Aa = pAx,$$

$$p = p'\frac{a}{x},$$

so that

$$W = Ap'a\int_a^l \frac{dx}{x}$$

$$= Ap'a\,\log_e\frac{l}{a}.$$

Note that $Ap'a$ is the work done on the piston during admission.

624. Find an expression for the *entire* effective work of the forward stroke of an engine working under the conditions above named except that there is a constant back pressure (condenser pressure) p_c.

Note that the pressure of admission is constant, as is also the back pressure. The work due to these pressures is readily calculated.

625. (*a*) Apply the results of the last problem to finding the work per forward stroke when the numerical data are :

    Area of piston   = 100 sq. in.
    Length of stroke   = 14 in.
    Boiler pressure   = 60 lb. per square inch.
    Back pressure   = 2.5 lb. per square inch (actual).
    Cut-off at $\frac{3}{14}$ stroke.

If an ordinary steam gauge shows 60 lb., the actual pressure is 60 + 14.7 lb. per square inch.

(*b*) The engine is double-acting and makes 180 revolutions per minute. Compute the horse-power.

626. As the result of an engine trial the data are :

Mean effective pressure from indicator card = 32.6 lb. per square inch.
Area of piston   = 64 sq. in.
Length of stroke   = 10 in.
Speed   = 400 revolutions per minute.

The indicator diagrams being the same on both sides of the piston, it is required to find the indicated horse-power.

627. Why are condensing engines more efficient than those which exhaust into the air ?

628. A perfect engine takes steam from a boiler at 150° C., and exhausts into a condenser at 30° C. Compute the efficiency.

629. If a compound marine engine consumes 2 lb. of coal per indicated horse-power every hour, what per cent of the energy of the coal is being transformed into work in the cylinder? The heat value of 1 lb. of coal may be taken at 12,000 B.T.U. (pound, degree Fahrenheit units).

ELECTRICITY

STATIC ELECTRICITY

630. Two bodies are rubbed together and then separated. It is found that they are electrified and have energy. What is the source of this energy?

631. Draw diagrams showing how an electric charge distributes itself over the surface of a conductor. What fundamental law of electrostatics explains this?

632. Two unit quantities of electricity are placed 10 cm. apart in air. What force will be exerted between them?

633. A charge of +10 is 25 cm. from a charge of −40. Find the force exerted between them.

634. The force between two charges is measured; each charge is then doubled. What will the force be if the distance is unchanged? How much must the distance between them be altered that the force may be as before?

635. The distance between two charges is 16. cm. One charge is +20. What must the other be in order that the force of repulsion may be 2 dynes?

636. Two charges q and q' are r cm. apart. q' is doubled, q divided by 8, and r is altered so as to leave the force unchanged. Find change in r.

637. Explain why light uncharged bodies are attracted when a charged body is brought near them.

638. Explain fully how a gold-leaf electroscope is charged by induction. State briefly how the lines of force are distributed at each step.

639. Define *surface density.*

640. A sphere of radius 20 cm. is charged with 400 units of electricity. What is the surface density?

641. The quantity on a sphere is increased fourfold. How must the radius be changed that the surface density may be the same?

642. What is meant by a line of force? a field of force?

643. A charge of 80 units is placed at a point where the strength of field is 100. What force will act on the charged body?

644. Would the presence of a field of electric force be observed if no charged body were placed in it?

645. Explain why an electrophorus may be used to obtain a considerable quantity of electrification with only a small initial charge.

646. An electrophorus (the lower plate) is charged. What will be the nature of the electrification of a body charged by means of it?

647. In using an electrophorus we may divide the process into four parts: (1) the approach of the metallic plate to the charged one; (2) "grounding" the upper plate; (3) separating the two; (4) the discharge of metallic plate.

648. Draw diagrams showing the distribution of charge in each case of Example 647.

649. Draw the lines of force in each case of Example 647.

650. Two equal light insulated conducting spheres are suspended so as to hang near together. One is charged positively. Will it attract the other? The second is grounded. Will the force action be altered, and how?

651. If instead of grounding the second they had been brought in contact and then separated, what change in the force action would be observed?

652. Give numerical values to the charge and distance between the centers of the spheres in the latter case, and find the force action before and after contact.

653. Define electrical potential at a point. In what units is it measured?

654. An isolated charge causes a potential of 25 at a point near it. What would the potential be if the charge were increased fourfold? if a charge opposite in sign, and twice as large, were combined with the first?

655. Show that for a single charge q the potential, at a distance r, is $\dfrac{q}{r}$.

$$V = \int_{\infty}^{r} q \frac{dr}{r^2}, \text{ etc.}$$

656. Find the potential at a point midway between A and B in Fig. 51; between B and C.

$AB = 1$ m., $BC = 20$ cm.

$q = 160$ $\qquad\qquad\qquad\qquad\qquad q' = -80$

A●---------------------------C---●B

Fig. 51.

657. How much work would be required to move a charge of $2 +$ units from a point on AB 10 cm. from A to one 10 cm. from B? (In Fig. 51.)

658. A small sphere is charged with $40 +$ units. Draw the distance-potential curve, taking the origin 1 cm. from center of the sphere $(r < 1$ cm.$)$. Draw the distance-force curve in the same way. Where do these curves intersect? How might the second be derived from the first?

659. A conductor 20 cm. long is placed in an electrical field. The potential at the points occupied by its ends would be 40 and 10 respectively, if the conductor were absent. How would the potential of these points be altered by the introduction of the conductor?

660. What takes place on the conductor when it is moved across the equipotential surfaces of the field?

661. Two spheres of equal radii are suspended by silk threads, and each is grounded. After the "ground" is broken charged bodies are brought in the neighborhood, such that the potentials at the points occupied by the center of the spheres would be at potentials 10 and − 10 respectively. What changes would occur if the spheres were connected by a wire?

662. A sphere of radius 10 is charged so that the surface density is $\frac{100}{4\pi}$. What *quantity* is required? What is the potential of a point just outside the sphere? What is the electric force at that point? Would any of these quantities be altered if the sphere were immersed in turpentine? Explain.

663. What work is done in moving a charge of + 30 from a point where $V = 40$ to one where $V = 100$?

664. To move a charge of + 4 from $V = -10$ to $V = +10$ will require how much work?

665. A small sphere has a charge of 8+ units. Draw six equipotential surfaces; three having $V < 1$, one $V = 1$, two $V > 1$.

666. Indicate, *briefly*, the change in these surfaces if a charge of − 4 were brought to a point 9 cm. from the first sphere.

667. A charged sphere A is brought near to an insulated conductor B.

Describe the electrical state of B (charge and potential):

Fig. 52.

 (a) When A is placed near B.
 (b) After grounding B.
 (c) When B is again insulated and A removed.
 (d) When B is again insulated and A brought nearer than before.

668. Draw the lines of force in each case of Ex. 667.

669. Draw the equipotential surfaces of Ex. 667.

670. Two equal charges are 80 cm. apart. If each charge is +40, what is the potential half-way between them? What is the force at that point?

671. Indicate the difference between the electrical condition at a point half-way between two charges when they are equal, and when they are equal but opposite (*i.e.* force and potential at the point).

672. A small charged sphere is lowered through an opening into a spherical conducting shell. Draw the lines of force and the equipotential surfaces in the following cases :

(*a*) When charged sphere is near center of the shell.

(*b*) When brought quite near one side of shell.

(*c*) After touching the inside of the shell.

673. Show that the potential inside a closed spherical shell is constant. What conclusion concerning the electric force within a shell follows from this?

674. A straight line is drawn in any direction across the lines of force and equipotential surfaces of a *uniform* field. What is the meaning of the ratio of the difference in potential between two points on the line to the distance between the points?

675. What is the meaning of the above ratio when the field is not uniform? when the field is variable, but the distance between the points is very small?

676. Assuming that the charge on an isolated sphere acts on a small charge just outside the sphere as though the entire charge were placed at the center, show that the electric force just outside is $4 \pi \rho$ ($\rho =$ surface density). Since *independent* of radius of the sphere, what follows in regard to an infinite charged plane?

677. Can two equipotential surfaces intersect? Can an equipotential surface intersect itself? Explain your answers.

678. Explain how an insulated conductor in the presence of charged bodies remains an equipotential region.

679. A charged sphere is placed between two very large conducting plates. Draw the lines of force and equipotential surfaces.

680. What peculiarity of the distribution of the lines of force indicates a strong field? of the equipotential surfaces?

681. Draw a curve showing the relation between the charge and potential of an isolated conductor, using Q as x and V as y. What does the slope of the line mean? What does the area of the curve mean?

682. After Q has reached a certain value, a grounded conductor is placed near the first, and Q is again increased. What changes in the QV line would indicate this?

683. When Q is stationary, and the second conductor is near, they are both surrounded by paraffine and Q is again increased. Show how the QV line would differ from the preceding.

684. A conducting sphere A is charged and placed on an insulating support at a great distance from all other conductors. Another conductor, B (uncharged), is brought near A. Will the charge on A be altered? the distribution? the potential of A? the force at neighboring points? If the distribution of force is altered, where would it be increased and where diminished? Answer the same questions if B were "grounded."

685. A straight line is drawn in any direction in a uniform field. If the potential at each point of the line be taken as y, and distances from a fixed point on the line as x, what kind of a curve will be found? What will the slope mean? What will the slope be when the given line is drawn perpendicular to the lines of force? When will the slope be a maximum?

686. Explain fully the difference between the electric force at a point, and the electric potential at that point. What relation is there between them?

687. Is potential a directed quantity or vector? Find the dimensions of electric potential.

688. The difference of potential between two points is 500; the distance between them is 40 cm. What is the average field strength between them?

689. The average field strength between two points is 50; they are 2 m. apart. What is the difference of potential?

690. Find the term C, V, or Q, omitted in the following table, where C = capacity of a conductor; V = the potential to which it is raised; Q = charge required to give a potential V.

Q	V	C
80		20
20		80
80	20	
20	80	
	80	20
	20	80

691. Find the energy in each case of Example 690.

692. What is meant by the term *capacity* as applied to a conductor or condenser?

693. A charge of 400 raises the potential of a sphere from 0 to 100. What is its radius?

694. Three spheres, capacities 4, 8, 12, respectively, are charged to potentials 24, 16, and 8. What is the quantity on each? The spheres are connected by a wire of negligible capacity. What will be the common potential?

695. What energy is required to charge a sphere of radius 10 to a potential of 100? of radius 100 to a potential of 10? to charge a sphere of radius 10 with a charge of 100?

696. The radius and charge on a sphere are each increased threefold. How is the potential affected? the energy?

697. (*a*) Upon what does the electrical *capacity* of a conductor depend? Explain why the capacity of a body is altered by bringing a grounded conductor near. (*b*) If a body whose capacity is 200 C.G.S. is charged to a potential of 4 (C.G.S.), what is the quantity of electricity? How much work is done in charging the body? (If formulas are needed *derive* them.) (Winter, '96.)

698. *A* and *B* are two spheres, radius of each 1 cm. What is the capacity of each?

699. *A* is given a charge of + 80. *B* is given a charge of − 40. The distance between their centers is 50 cm. Locate a point on the line joining their centers where $V = 0$; $+ 2$; $- 3$.

700. *B* is brought in contact with *A* and then replaced. How would the charges be altered? What change in potential would occur at each of the points mentioned above?

701. What do you mean by a condenser? Upon what does the capacity of a condenser depend?

NOTE. — Unless otherwise stated, it will be assumed that one coating of a *condenser* is grounded, *i.e.* $V = 0$.

702. State the analogy between electric condensers and water reservoirs.

703. A condenser of capacity 1000 is charged with 500 units. Half of this charge escapes. What proportion of the energy has been lost?

704. A quantity *Q* charges a condenser to a potential *V*. What energy is stored?

705. The area of the plates, the thickness of the dielectric and its specific inductive capacity are each doubled. How will its capacity be changed?

706. Define specific inductive capacity.

707. A certain condenser when air is used as the dielectric has a capacity of 400; when glass is substituted, the capacity is found to be 2600. What is the specific inductive capacity of the glass?

708. The *force action* between two charged plates is found to be one-third as great when shellac is between them as when air is the dielectric. Find the specific inductive capacity of shellac.

709. Derive the formula for the capacity of a spherical condenser: radii of conductors r_1 and r_2, specific inductive capacity of dielectric k.

710. Derive the expression for the energy required to charge a condenser in terms of Q and V; in terms of Q and C; in terms of C and V.

NOTE.—$dW = VdQ$. But V is a function of Q, $V = \dfrac{Q}{C}$.

$$\therefore\ W = \int_0^Q \frac{1}{C} QdQ,\ \text{etc.}$$

711. Compare the energy required to charge two spherical condensers to the same potential when the radii of the shells of one are 20 cm. and 20.1 cm., sp. ind. cap. of dielectric 2, while for the other these quantities are 40, 40.2, and 6.

712. A condenser of capacity 50 and charge 400 is connected by a poor conductor to earth until its energy is reduced to one-sixteenth of its initial energy. What charge escapes? How much is the potential decreased?

713. It is observed that the energies of discharge of two jars charged from the same source to earth are as 1 to 9. Find the ratio of their capacities.

714. *A* and *B* are two reservoirs of the same volume, but of unequal height. *P* is a pump powerful enough to force water to the top of *A*.

(*a*) Which would possess the more potential energy when filled?

(*b*) Which would exert the greater pressure when full?

(*c*) The stop-cock k is closed when *B* is full, and *A* is filled, k' is closed, and k is opened. What change in energy distribution occurs?

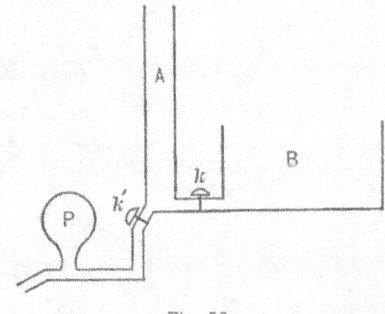

Fig. 53.

(*d*) If the system were connected with a reservoir *below* the

K

level of the source from which the water is pumped, how would the available energy be altered?

State the analogous electrical problem for each case.

715. Draw a diagram of a charged Leyden jar when one coating is grounded, showing the distribution of lines of force and equipotential surfaces.

716. Two oppositely charged and insulated plates are placed parallel to each other and near together. Explain why when either is touched only a slight shock is received.

717. Would an increase of the distance apart change the effect, and if so, how?

718. What effect would an increase of the specific inductive capacity of the medium between the plates have?

719. There are three conducting spheres of equal radii. The first is charged and brought in contact with the second, this in turn brought in contact with the third. Find the energy changes in each operation. How much energy is still stored in the system? How much was stored in the first sphere? What relation exists between these quantities?

720. What would be the capacity of a plate condenser when the area of each plate is 1 sq. m., the distance apart is .1 cm., the specific inductive capacity of the dielectric being 4?

721. How much energy is required to charge such a condenser to a potential of 100?

722. In the discharge of a condenser what becomes of the energy? What experiments confirm your statement?

723. How would you proceed in order to charge a Leyden jar?

724. Find the energy of discharge of a condenser when the plates are of potentials V_1 and V_2, and the capacity is C.

725. There are three Leyden jars, A, B, and C, equal in capacity, having their outer coatings connected to earth. A is

first charged.   Its knob is then connected with the knob of B.
It is then disconnected from B and connected with C.   Finally
the knobs of A, B, and C are connected.   Find the energies
of the several discharges.   (Larden.)

726. When are two or more condensers said to be connected
in " series " ?   When in parallel or multiple ?

727. The inner plates of four similar condensers are joined,
and each outer plate is grounded.   What is the ratio of the
capacity of the set to that of a single one ?

Compare : (a) The potential to which a given charge would
raise the system with that to which it would raise one.   (b) The
energy required to raise the system to a given V with that
required for one ?

728. Four similar condensers are joined in series ; the outer
plate of the last is grounded, the inner plate of the first is
charged to a potential V.   The capacity of each condenser is
C.   What is the potential of each jar ?   What is the total
charge ?   What is the entire energy stored ?

729. Two spheres, A and B, radii 5 and 2 respectively, and
charges $+ 40$ and $- 10$ are joined by a wire of negligible
capacity.   Find the capacity of the system ; the quantity on
each sphere ; the amount of electricity which has passed along
the wire ; the initial energy and the final energy.

CURRENT ELECTRICITY

730. State Ohm's law. For what kind of conductors and under what conditions is it true?

The units used in measuring current, electromotive force or potential difference, and resistance are named the **ampere, volt,** and **ohm** respectively. The relation of these to the C.G.S. system will be illustrated later (see p. 187). Ohm's Law is not dependent on the units employed. Hence in any system $I = \dfrac{\text{potential difference}}{\text{resistance}}$. In the practical system, current in *amperes*

$$= \frac{\text{potential difference in } volts}{\text{resistance in } ohms}$$

731. When potential difference = 80 volts, resistance = 40 ohms, what current will flow? What quantity will pass each cross-section of the wire in 5 min. (1 coulomb = 1 ampere second).

732. The terminals of a wire of 10 ohms' resistance are at potentials + 40 and − 40 respectively. What is the current strength : when at + 60 and − 20? when at 80 and 0?

733. The potential at each end of a circuit is multiplied by three. How must the resistance be changed that the current may remain the same?

734. A quantity of 200 coulombs is transferred along a wire in 40 sec. What is the current strength?

735. A current of strength 40 continues 2 min. What quantity passes?

736. *A* and *B* are two charged conductors. V_A is +, V_B is −. They are connected by a poor conductor. What changes of potential will take place?

737. In the above case, if the charge on A is reduced 80 + units and the charge on B is reduced 80− units, what is the total quantity which has passed along the connecting wire?

738. If this transfer takes place in 5 sec., what is the average current strength?

739. Two bodies of different potential are joined by a moist thread. It is observed that the change of potential is slow and the current is small. Explain.

740. What do you mean by the resistance of a conductor? What effect does the resistance of a conductor joining two points of constant difference of potential have?

741. Find the terms omitted, I, potential difference, or R, in the following table:

Potential Difference.	R	I
120	5	
.5	200	
500	250	
25		5
115		20
340		17
	35	7
	400	50
	2000	.0005

742. The terminals of a wire of resistance 60 ohms are kept at potentials of 100 and 10 for 5 min.; the terminal of lower potential is then "grounded" and the potential of the other reduced to 90; current flows again for 10 min. Compare the quantities transferred.

743. If in the equation $V = I \cdot R$, we take each quantity in turn as constant and the others as x and y, what loci would be obtained?

744. A uniform wire AB is kept at a uniform temperature, and its ends at a constant difference of potential. Draw a diagram showing the relation between the fall of potential and length of the wire.

745. If in Example 744 $V_A - V_B = 100$ volts, and $V_A = 200$ volts, what will be the potential midway between A and B? at one-fourth the distance from A to B?

746. The electromotive force of a battery is 4 volts, and its resistance is 6 ohms. The external resistance consists of four pieces of wire in series; their resistances are 10, 20, 30, and 40 ohms, respectively. Find

(*a*) the total current,

(*b*) the fall of potential along each wire,

(*c*) the difference of potential of the terminals of the battery.

747. Explain the difference between electromotive force and difference of potential.

748. A Leclanché cell is connected in series with a low-resistance galvanometer. The deflection of the galvanometer is observed to steadily decrease. Give two causes which may explain this.

749. If the cell is shaken, the deflection rises to nearly its former value. Explain.

750. (*a*) What is meant by *polarization* in the case of a galvanic cell? (*b*) Explain the action of some cell in which polarization is prevented.

The relation between current, potential difference, and resistance throughout a circuit may often be best understood by a properly constructed diagram. We may choose either of two ways, according to the end in view. We may assume any potential we please as our arbitrary, o, since we are concerned only with differences of potential. Then V may be plotted as y and R as x, or we may use V as y and *distances* measured from an arbitrary point in circuit as x. In case the circuit contains sources of electromotive force, we may usually consider the rise of potential through them as sudden, and the line becomes a broken one. If, however, the source of electromotive force is distributed like the armature of a dynamo, the line in such places would be curved. Potential-resistance curves are of considerable importance, and the student is advised to study carefully the simpler cases explained below before drawing those of more complicated circuits.

Take the case of a single cell, electromotive force 3 volts, internal resistance 6 ohms, external resistance 10 ohms. Starting at any point as B, and

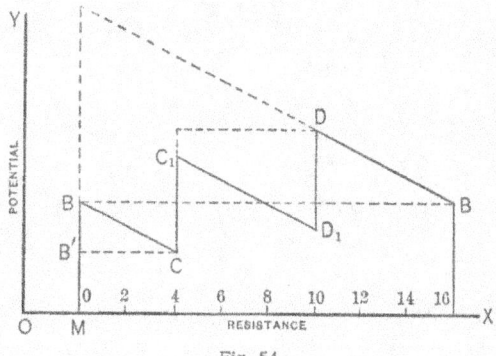

Fig. 54.

assuming MB as representing the potential at B, Ohm's law states that along BC the potential falls uniformly, so that $\dfrac{BB'}{B'C} = I$.

At C we may suppose an abrupt rise of potential taking place at the bounding surface of liquid and plate, then another uniform fall due to the resistance of the cell, another rise at D_1 falling again along DB to the value MB. Note that the lines of fall are all parallel, which is equivalent to the statement that the current is the same throughout the circuit. Suppose now that the external resistance were increased, I must decrease, and all of the sloping lines would become more nearly parallel with OX. But the vertical lines CC_1 and DD_1 are constant in length and independent of R; it follows then, in order that C_1D may remain parallel to BC and D_1B, either CC_1 must fall or DD_1 rise, or both. This is the same as saying that the *difference of potential of the terminals of a battery depends upon the external resistance, and approaches the electromotive force of the cell as this resistance is increased.*

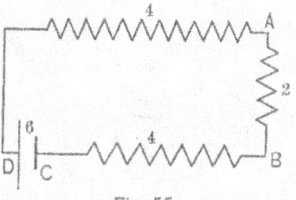

Fig. 55.

When potential and distance from a fixed point are used as co-ordinates, the lines of fall would not be uniform in slope, and the diagram would show through what absolute lengths of the circuit the fall is greatest.

The relations between current, resistance, electromotive force, and potential difference may often be better understood by reference to the flow of water in pipes, in so far as the analogy between the two exists.

In Fig. 56 suppose P is a pump capable of forcing water to a height H_0, connected to a tank T, from which leads a straight pipe $A_1A_2 \cdots S$; A_2H_2, A_3H_3, etc., a series of vertical pipes opening from the main whereby the pressure at each point can be measured; S a stop-cock whereby the flow in the main can be checked. When S is open and the pump working, so that

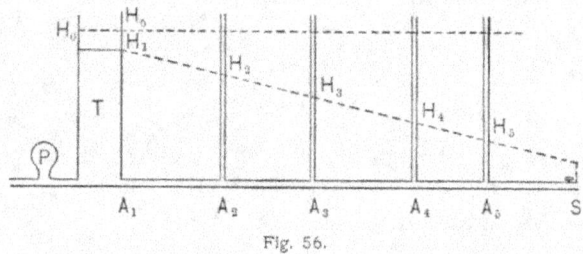

Fig. 56.

the current is *steady*, the pump will be unable to keep T full up to H_0, and it will be found that the tops of the water columns will be in a straight line $H_1 H_2 H_3 \cdots$.

751. What is the electrical analogue of :

    (*a*) The friction of the pipe ?

    (*b*) The friction of the pump ?

    (*c*) The pressure at A_1 ?

    (*d*) The difference between $A_1 H_0$ and $A_1 H_1$?

    (*e*) The ratio,

$$\frac{\text{pump pressure}}{\text{total friction}} = \frac{\text{difference of pressure between } A_1 \text{ and } A_2}{\text{friction between } A_1 \text{ and } A_2} ?$$

    (*f*) The height of line $H_1 H_5$ vertically above S ?

    (*g*) Current and quantity ?

    (*h*) The changes which occur when S is slowly closed ?

752. Would the analogy hold if the pipe were bent ? if it were enlarged at some point ?

753. State a case in flow of water analogous to cells in series ; in multiple. Explain fully.

754. In the circuit shown (Fig. 55), a point in the external resistance is " grounded." Draw the potential-resistance curve. What change in your diagram would indicate a change in the position of the ground ?

755. Determine what external resistance is required in the

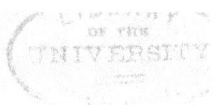

circuit of Fig. 57 in order that the potential difference of A and
B may be 1 volt? $1\frac{1}{2}$ volts? $\frac{1}{4}$ volt?

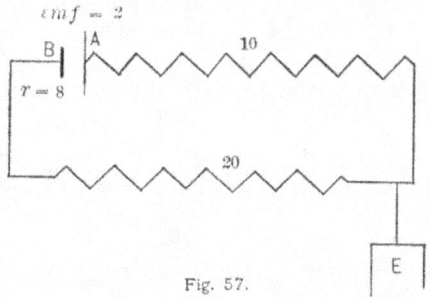

Fig. 57.

756. If the resistances of the cells in Fig. 58 are very small,
draw the potential-resistance curve.

SUGGESTION. — Each electromotive force causes a rise of V independent of
the other.

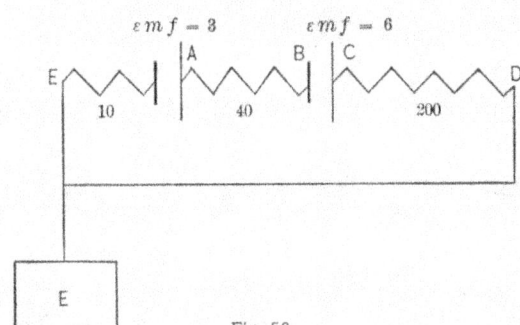

Fig. 58.

757. What is the potential difference between A and B,
Fig. 58?

758. The electromotive force of a battery is 5; when the
external resistance is 100, the potential difference at the terminals is 4.   What is the internal resistance?

759. A circuit consists of three cells, in series; E.M.F.'s 1,
2, 3; resistances 4, 5, 6, respectively.   The external resistance
is 20 ohms.   Draw the potential-resistance curve.   What is the
potential difference between the negative plate of the first and
the positive plate of the last?

760. In a conductor where the resistance increases as the square of the distance from the end (decreasing cross-section), draw a curve, using V as y, and distance from one end as x, when the potential difference of ends remains constant.

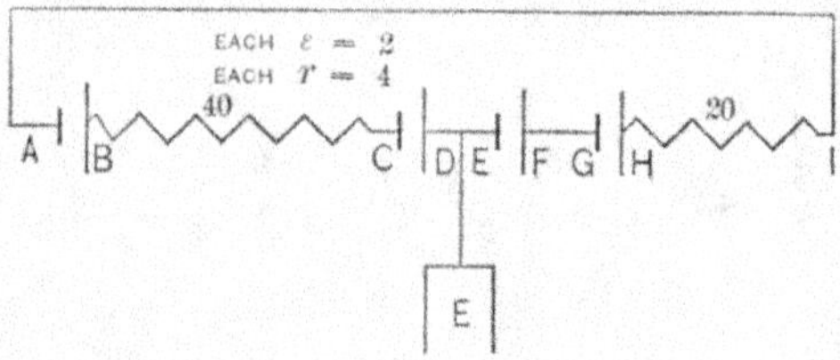

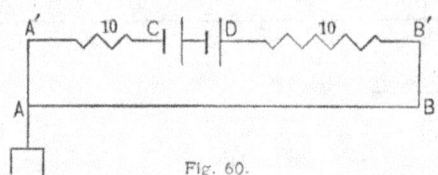

EACH $\varepsilon = 2$
EACH $r = 4$

Fig. 59.

761. Draw the potential-resistance curve for the circuit in Fig. 59 :

         (*a*) When "ground" is broken.

         (*b*) When "grounded" as shown.

762. Each cell in Fig. 60 has an electromotive force of 2 volts, and a resistance of .4 ohms. Other resistances as shown. All connecting wires ($A'A$, AB, etc.) are so large that their resistance can be neglected. A is connected to the earth.

Fig. 60.

(*a*) Draw diagram to show the variation of potential along $A'CDB'$. (*b*) Compute the difference of potential between C and D.

763. Name four things upon which the resistance of a wire depends.

764. Two copper wires are of the same cross-section, but one is twice as long as the other. Compare their resistances.

765. What do you mean by the resistance of wires in *multiple* or *parallel?*

766. How is Ohm's law applied to find how *multiple* resistance depends on the resistance of the separate branches?

767. The length of a wire is increased fourfold. How much must its radius be changed that its resistance may be the same as before?

768. An iron wire of a certain length and cross-section has a resistance of 40 ohms. What would be the resistance of an iron wire ten times as long and one-fifth the diameter of the first?

769. What would be the resistance of n equal resistances joined in multiple? in series?

770. Thirty incandescent lamps, each $R = 50$ ohms, are joined in multiple. What is their combined resistance?

771. Find the resistance between two points in a circuit when they are joined by:

(*a*) Three wires in *multiple*, resistances 2, 5, 7, respectively.

(*b*) Three wires in series, resistances 2, 5, 7, respectively.

(*c*) Four wires in multiple, resistances 40, 20, 30, 50, respectively.

(*d*) Four wires in series, resistances 40, 30, 20, 50, respectively.

772. The resistance between two points in a circuit is 60 ohms. What must be placed in multiple with this to reduce the resistance to 22 ohms?

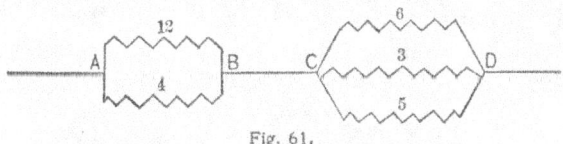

Fig. 61.

773. What is the resistance between A and B? C and D? A and D? Fig. 61.

$$\frac{1}{^AR_B} = \frac{1}{4} + \frac{1}{12} = \frac{4}{12}, \text{ or } {}^AR_B = \frac{12}{4} = 3 \text{ ohms.}$$

774. A copper wire of length l is divided in the ratio of 3 to 5, and the pieces joined in multiple. What *length* of the same wire might have been taken to get the same resistance?

In dealing with a complex circuit it is well to compute each multiple resistance first, and then deal with the set in series.

775. Find the total resistance of the circuit, Fig. 62. In this system we may compute the resistance from A to B, then from C to D, finally add together all the resistances in series.

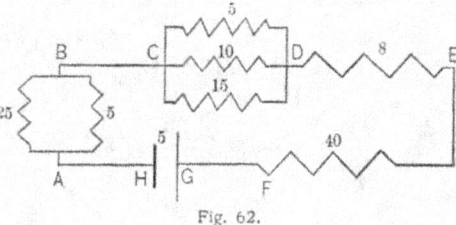

Fig. 62.

776. Find the total resistance of the circuit, Fig. 63. (Compute each multiple resistance first.)

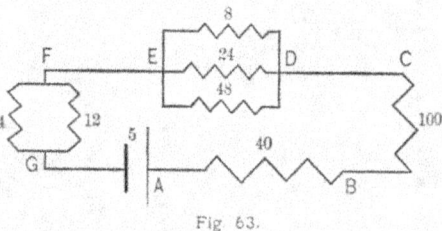

Fig 63.

777. Find the resistance of the system shown in Fig. 64.

778. AC and BD (Fig. 65) are two metal plates of 0 resistance. A and B are joined by a wire of 10000 ohms resistance. Find x_1, so that when placed in multiple with the first the combined resistance is 1000. Then x_2, so that multiple resistance of the three is 100, etc.

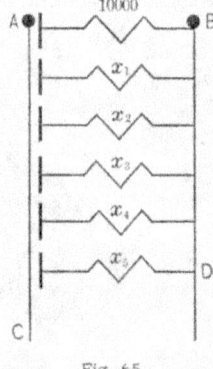

Fig. 65.

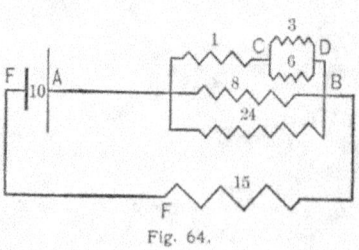

Fig. 64.

779. Prove that the resistance of two wires in multiple is always less than that of either.

780. Prove the following construction for computing multiple resistance.

Lay off on OY a length r_2.

Lay off any line ‖ to OY a length r_1.

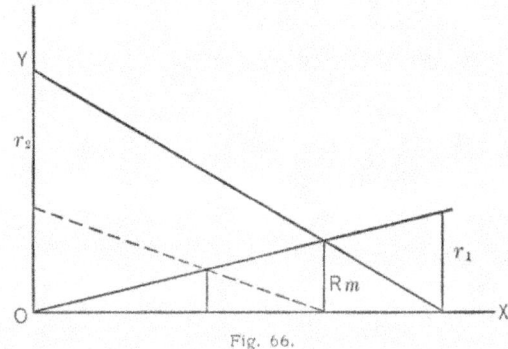

Fig. 66.

Join the upper end of each line with the lower of the other. The ordinate of the intersection of these lines is the resistance required.

For three or more resistances we may extend the construction as for r_3. By using cross-section paper the results may be quickly obtained.

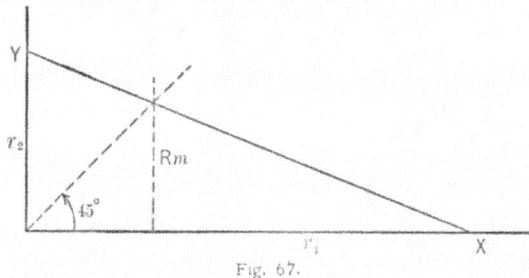

Fig. 67.

781. Prove the following construction for resistances in multiple. Take $x = r_1$, $y = r_2$; join their extremities. Then the resistance of r_1 and r_2 in multiple is given by the co-ordinate (y or x) of the intersection of this line, with a line drawn at an angle of $45°$ with the axes.

This may be extended to any number of resistances in multiple, and easily effected by the use of cross-section paper.

In the following problems it should be remembered that in dealing with cells in multiple and in series we must be careful to consider *both* the electromotive force and the resistance of the combination. It is *assumed* that the cells are exactly alike, both in resistance and electromotive force, unless otherwise stated.

The electromotive force of any number (n) of cells in *series*
         = sum of electromotive forces.
The electromotive force of any number (n) of cells in *multiple*
         = electromotive force of one.
The resistance of any number (n) of cells in *series*
         = sum of resistances.
The resistance of any number (n) of cells in *multiple* is computed just as any other multiple resistance.

782. Six cells, resistance of each 12 ohms, electromotive force of each 2 volts are connected in series. Find combined electromotive force and resistance. Find them when in multiple.

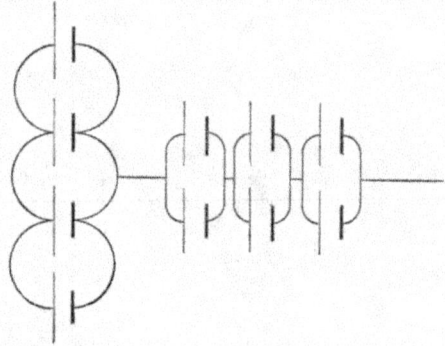

Fig. 68.

783. A system of ten cells, electromotive force 3 volts, r 6 ohms, are connected as shown in Fig. 68. Find the electromotive force and resistance.

784. A system of fifty cells, electromotive force 1 volt, r .4 ohms, are placed "ten in a row" (series), and the five rows in multiple. What is the internal resistance of the battery? the electromotive force?

785. Find the current strength when each circuit (Examples 783 and 784) is closed by an external resistance of 200 ohms.

786. Given twenty-four cells, electromotive force 2 volts, r 4 ohms, external resistance 5 ohms. Separate 24 into its various factors (as 2, 12; 3, 8; etc.); choose each factor in turn as the number of cells in a row, and the other as the number of rows. Compute the current strength in each case.

787. Do the same when external resistance = 1 ohm; 200 ohms.

788. When two or more *wires* are joined in multiple, at each junction they have a common potential. Hence by Ohm's law the current through any wire will be the common potential difference between A and B divided by the R of that wire.

789. Three wires in multiple (Fig. 69); potential difference between A and B = 24 volts; resistances as shown. What current flows in each branch? What is the *total* current?

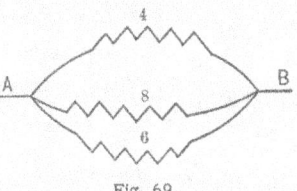

Fig. 69.

790. The currents in two branches of a divided circuit are as 4 to 12. What is the ratio of their resistances?

791. In the circuit shown in Fig. 70, find

    (*a*) The total electromotive force.

    (*b*) The total resistance.

    (*c*) The total current.

    (*d*) The fall of potential between K and G.

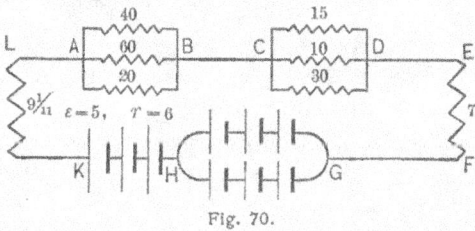

Fig. 70.

    (*e*) The fall of potential between A and B; C and D; E and F.

    (*f*) The current in each branch between A and B.

State your reason for each step in the numerical work.

792. Twenty 50-volt lamps, each requiring 1.2 amperes, are connected as shown. The resistance of BB' and CC' is nearly o, that of $AB + DC$ is 1 ohm.

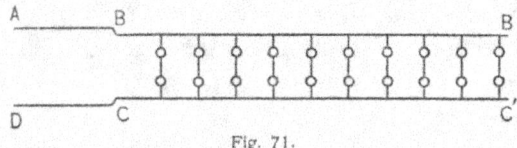

Fig. 71.

Find (a) The resistance between B and C.

(b) The total current.

(c) Difference of potential between B and C.

(d) Difference of potential between A and D.

(e) The heat developed per minute in the lamps.

(f) What change takes place when five pairs of lamps are turned off ?

(g) What objection would there be to short-circuiting *one* of each *pair* of lamps ?

793. A resistance of 80 ohms joins the terminals of a battery, electromotive force 100, resistance 20. A shunt of 5 ohms is placed around 20 ohms of the external resistance. What effect will this have on the total current ? What effect on difference of potential of the points where it is joined ?

794. In what case will a shunt placed around a portion of a circuit have no appreciable effect on the total current ?

795. State and explain Ohm's law. If the connections and resistances of a certain circuit are as shown in Fig. 72, compute the current flowing in each of the two branches between A and B. Each cell has an electromotive force of ·1 volt and a resistance of 5 ohms.

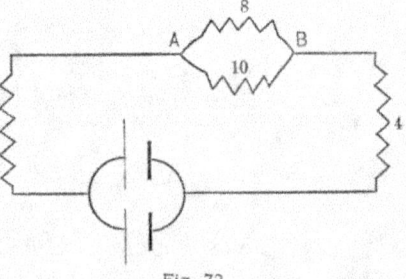

Fig. 72.

796. The resistance between A and B is 100 ohms. What resistance, x_1, must be placed in shunt with this in order only .1 as much current will flow along AB as before? (V_A potential difference, V_B to remain the same.) Find x_2 so as to reduce the current in AB to .01 of its former value, etc.

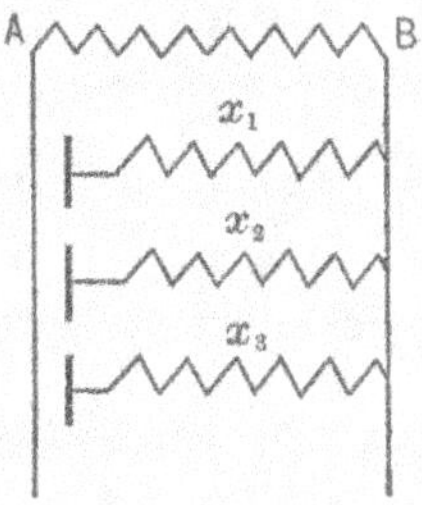

Fig. 73.

By Ohm's law, $x = 100 \cdot I_R = x_1 \cdot I_x.$

But $I_x = 9 I_R.$

$\therefore\ 100\, I_R = x_1\, 9\, I_R.$

$\therefore\qquad x_1 = \tfrac{100}{9} = 11\tfrac{1}{9}$ ohms.

(Compare Example 778.)

797. Prove that when a shunt of resistance s is placed around a wire of resistance r the current is $r = \dfrac{s}{s + r} \cdot$ total current. Extend this to three or more resistances in multiple.

$$\left(\text{In general, } I_{r_1} = \frac{r_2 \cdot r_3 \cdot r_4 \cdots r_n}{\Sigma r_1 r_2 \cdots r_{n-1}}.\right)$$

798. A galvanometer of 1980 ohms resistance is "shunted" by a wire $r = 20$. What proportion of the total current passes through the galvanometer?

799. The difference of potential between A and B, Fig. 63, is to be measured by placing an instrument (voltmeter) in shunt with the resistance between A and B. What change in this difference of potential is caused by the insertion of the instrument?

800. In the circuit of Fig. 70, what is the *smallest* resistance a voltmeter could have that when placed in shunt with AB the difference of potential between A and B may be changed only one-half of one per cent?

801. The current between two points in a circuit is to be measured by passing it through a measuring instrument (ammeter). Under what conditions is the current unaltered by the introduction of the ammeter?

L

802. In the circuit shown in Fig. 70, what is the largest resistance which an ammeter could have and only alter the current strength one-half of one per cent?

803. APB and AQB (Fig. 74) are two conductors joined in multiple. A and B are kept at different potentials. Draw the potential-resistance diagram for each path from A to B.

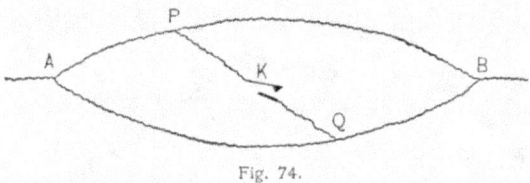

Fig. 74.

804. If potential at A is 50 volts, and at B is 40 volts, what range of potentials may be found along APB? along AQB?

805. If P and Q are two points of the *same* potential and the key k is closed, would the distribution of the potential be altered?

806. When is it certain that if any point P is chosen on the upper branch, a point of the same potential can be found on the lower one? Explain fully.

807. If a source of electromotive force were in any part of the circuit between A and B, would it always be possible to find for any potential along APB a corresponding point in AQB?

808. Find the relation between the resistances AP, PB, etc., when $V_P = V_Q$, in case of no electromotive force between A and B. (Wheatstone's bridge.)

809. Show that the best arrangement of a given number of cells is that which makes the external and internal resistances as nearly equal as possible.

$$I = \frac{nE}{\dfrac{nr}{m} + R}$$

$$= \frac{mnE}{nr + mR}.$$

[E = electromotive force of one cell.
[r = resistance of one cell.
[R = external resistance.
[n = No. cells in a "row."
[m = No. of rows.

Since mn = number of cells, the numerator is constant.

$\therefore$ I is a maximum when $nr + mR$ is a minimum;

i.e. $nr + mR$ is a minimum by variation of n and m.

$\therefore$ $rdn + Rdm = 0.$

But $\qquad\qquad mn = $ constant.

$\therefore$ $mdn + ndm = 0.$

Whence $\qquad\qquad \dfrac{r}{m} = \dfrac{R}{n}$ or $\dfrac{nr}{m} = R.$

It does not follow that the two simultaneous equations $mn = N$ and $\dfrac{nr}{m} = R$ have *integer* roots; and as fractional parts of a cell are meaningless, we must choose the two factors of N which make $\dfrac{nr}{m}$ as nearly R as possible.

810. Deduce from the statement of how to group for maximum current a rule when the external resistance is very great ; very small.

811. How would you group twenty-four cells, each $r = 6$, $E = 3$, $R = 16$, for a maximum current ? $R = 36$? $R = 9$? $R = 25$?

$$\frac{n \cdot 6}{m} = 16,$$

$$mn = 24.$$

Multiply these equations,

$$n^2 \cdot 6 = 16 \cdot 24,$$

$$n^2 = 16 \cdot 4 \text{ or } n = 8.$$

$$\therefore m = 3. \qquad\qquad [\text{8 in a row, 3 rows.}$$

812. Apply Kirchhoff's laws to the circuit shown in Fig. 75, where electromotive force of the cell is E and the resistance of cell and connecting wires is r.

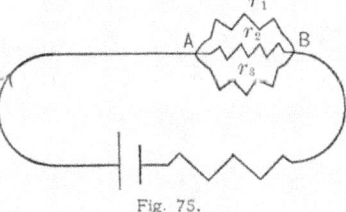

Fig. 75.

These laws are often stated as follows :

(1) If any number of conductors meet in a point $\Sigma I = 0$; or there is no accumulation of electrification at the point.

(2) In any complete circuit

$$\Sigma IR = \Sigma e.$$

In applying the first law, if we consider the current flowing *toward A* as $+$, we must consider those from A as $-$. While in the use of the second law, if we start from A toward B, *i.e. with* the current, and call $I_1r_1 +$, we must, when returning along r_2, take I_2r_2 as $-$.

By   (1) $I = I_1 + I_2 + I_3$.

           (2) $I_1r_1 - I_2r_2 = 0$,

              $I_1r_1 - I_3r_3 = 0$,

              $I_2r_2 - I_3r_3 = 0$,

              $Ir + I_1r_1 = E$.   [Where E = electromotive force of cell.]

From the first and second of (2) we may express I_2 and I_3 in terms of I_1, r_1, r_2, and r_3.

Substituting in (1),

$$I = I_1 + I_1\frac{r_1}{r_2} + I_1\frac{r_1}{r_3}$$

$$= I_1\left[\frac{r_2r_3 + r_1r_3 + r_1r_2}{r_2r_3}\right].$$

$$\therefore I_1 = \frac{r_2r_3}{r_1r_2 + r_1r_3 + r_2r_3}.$$ [Sim. for I_2 and I_3.

If R = equivalent resistance of r_1, r_2 and r_3,

$$IR + Ir = E,$$
$$I_1r_1 + Ir = E,$$
$$I_2r_2 + Ir = E,$$
$$I_3r_3 + Ir = E.$$

Add last three and equate to three times the first. Solve for R, using I_1 above.

$$R = \frac{r_1r_2r_3}{r_1r_2 + r_1r_3 + r_2r_3}.$$

813. Find the distribution of current in a set of five unequal resistances joined in multiple.

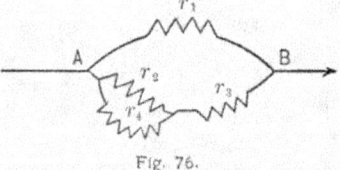

Fig. 76.

814. In the circuit of Fig. 76, show that

$$I\left(1 - \frac{r_1r_3}{r_3 + r_1^2}\right) = I_4\left(1 + \frac{r_4}{r_2} + \frac{r_4}{r_1} - \frac{r_4r_3}{r_3 + r_1^2}\right),$$

where I = total current.

815. The resistance of ADB is 10, of ACB is 40. Find the current in AB.

Assuming direction of currents as indicated by the arrows,

$$I_1 = I_2 + I_3,$$
$$I_3 \cdot 40 - I_2 \cdot 20 = 10,$$
$$I_1 \cdot 10 + I_2 \cdot 20 = 2.$$

Eliminating I_1 and I_3, we have $I_2 = -\frac{1}{70}$ amperes. What does the negative sign mean? Solve when the arrow from A to B is reversed.

816. Find I_2 when one cell is reversed (Fig. 77).

817. What electromotive force must be inserted in branch (1) (Fig. 77), that no current shall pass through (2)?

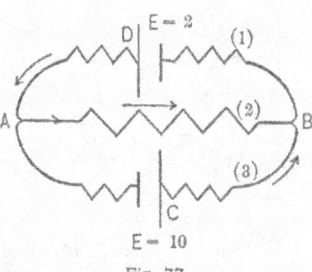

Fig. 77.

Put $I_2 = 0$, and the third equation $= e$.
Whence $E = 2\frac{1}{2}$ volts.

818. Three cells, electromotive force E_1, E_2, E_3, internal resistances r_1, r_2, r_3, are joined in multiple and the external resistance is R. Find the total current. Test your answer by reference to the case when the cells are alike.

$$I = \frac{e_1 r_2 r_3 + e_2 r_1 r_3 + e_3 r_1 r_2}{r_1 r_2 r_3 + R (r_1 r_2 + r_1 r_3 + r_2 r_3)}. \quad Ans.$$

819. Assume, in Example 818, $E_1 = 2$, $E_2 = 4$, $E_3 = 6$, $r_1 = 3$, $r_2 = 6$, $r_3 = 12$, $R = 40$. Find the current in amperes.

820. A and B are two points in a circuit which is carrying a current of 10 amperes. $^A R_B = 100$ ohms. What work is done in this portion of the circuit per *minute*? What becomes of this energy?

821. How much heat is developed per second in a portion of a circuit, potential difference of the ends 50 volts, and the current 50 amperes?

Current in amperes × potential difference in volts = energy in watts.

Heat per second in calories $= \dfrac{\text{watts}}{4.2} = \text{watts} \times .24$.

Or $H = I \cdot V \cdot .24 = I^2 R \cdot .24$.

822. The resistance of a conductor is doubled and the current halved. How is the heat developed affected?

823. The current in a wire is multiplied by three. How much must the resistance of the conductor be altered that the loss by heat shall be unchanged?

824. A current of 10 amperes develops 144.10^4 calories per minute. What was the resistance? What quantity passes per minute? What potential difference is required to maintain the current?

825. State clearly the meaning of the terms *watt* and *joule*. *Watts* × *time* = ?

826. A current of 40 amperes flowing in a coil causes a difference of potential of 20 volts between its terminals;

      (a) How much energy is consumed in 1 hour?

      (b) How much heat is developed?

827. Four wires of equal length and diameter, but of different specific resistances, are joined in series. For example, soft steel, copper, platinum, and silver are used. Find the ratios of the heat developed in the wires.

828. Given mn similar cells, each E.M.F. $= e$, resistance of each $= r$; external resistance R. How must they be arranged to secure the greatest heating effect?

829. A wire of resistance 1000 ohms is found to develop heat enough in 10 sec. to raise 24 kg. of water 10°. What current does the wire carry? What difference of potential was required to maintain it?

830. If work is done by the current in addition to overcoming resistance, would IE and I^2R have the same value? Explain.

831 Find the distribution of heat in the circuit shown in Fig. 72, when there is no back electromotive force.

832. When a given set of generators are connected so as to give a maximum current through a given external resistance, show that one-half the total heat is developed in the generator.

833. Three copper wires of equal length, diameters .1 mm., .3 mm., .5 mm., respectively, are joined in multiple. The electromotive force of the junctions is kept constant. Find the ratio of the heats developed in the wires.

834. Why are large conductors usually used to transmit electrical energy? Why is copper used in many cases rather than iron? What determines which shall be used?

835. Why is it desirable to transmit electrical energy at high potential?

836. Why is it desirable to transform a small current at high potential to a larger current at lower potential at the point where it is used?

837. A current of 40 amperes is sent over a line of 10 ohms resistance. What is the fall of potential in the line? If the end of higher potential is at $V = 1000$, what energy per second is delivered at the end of lower potential? What is the heat loss per second? Answer the last two questions if V at the higher end were 2000 volts.

838. The voltage at which a certain amount of power is supplied to a line is doubled. What is the effect on the heat losses? How much could the length of the line be increased and still have no more loss in the line than at the lower voltage? How might the cross-section of the wire be changed in order that, the length remaining the same, the heat loss is the same as at the lower voltage?

839. What considerations limit the voltage used in practical work?

In order to compare resistances of various substances as well as to compute the resistance of a conductor from its dimensions, it is convenient to know the resistance of a cube of the substance of 1 cm. edge, at 0°. The actual resistance depends somewhat on the purity and previous history of the specimen, so the values given either refer to pure specimens, or are average values. The resistance of such a cube is named the *specific resistance* of the material. The statement that the specific resistance of copper is $17 \cdot 10^{-7}$ means that

1 cm. length of a piece of copper 1 sq. cm. cross-section has a resistance of .0000017 ohms at a temperature of 0°.

The values of specific resistance used are taken from Landolt and Born-stein's *Physikalisch Chemische Tabellen*.

To find the resistance of a copper wire 10 m. long, 1 sq. mm. cross-section at 0° we have

$$R = \frac{17 \cdot 10^{-7} \cdot 10^8}{10^{-2}} = .17 \text{ ohm.}$$

840. The specific resistance of silver is $15 \cdot 10^{-7}$. Find the resistance of a silver wire 1 ft. long and $\frac{1}{1000}$ in. in diameter.

841. A copper wire of known resistance is to be replaced by a platinum wire of half the cross-section. What length must be chosen to have the same resistance?

842. Find the resistances of the following circular wires at 0°.

Material.	Length.	Radius.	Specific Resistance.
Hard steel	10 m.	.5 mm.	$314 \cdot 10^{-7}$
Soft steel	10 m.	.5 mm.	$157 \cdot 10^{-7}$
Copper	1 km.	.2 mm.	$17 \cdot 10^{-7}$
Platinum	100 m.	.2 mm.	$135 \cdot 10^{-7}$
Silver	100 m.	.2 mm.	$15 \cdot 10^{-7}$
German silver	100 m.	.2 mm.	$236 \cdot 10^{-7}$
Carbon	1 m.	.1 mm.	$59350 \cdot 10^{-7}$

843. From the table of specific resistances above, compute the resistances of wires 1 m. long and 1 sq. mm. cross-section in each case.

844. A wire is drawn out into an extremely long circular cone. If its radius at each point is a times the distance from the end, and the specific resistance of the metal is $35 \cdot 10^{-7}$, find the resistance of the wire.

Form the expression for the resistance of a length dl and integrate.

As a first approximation, and between certain limits of temperature, the change of resistance of a wire with temperature may be expressed as a certain percentage of the resistance at 0° times the temperature above 0. The statement that the temperature coefficient of copper is .00388 means that for each degree a copper wire is heated above 0°, its resistance is increased the .00388th part of its resistance at 0°.

845. The resistance of a coil of copper wire at 0° is 1785 ohms. What will it be at 40°?

The increase is .00388 · 40 · 1785.

$$\therefore R_{40} = 1785 \,[1 + .1552], \text{ etc.}$$

846. The resistance of an iron wire at 20° C. is 1010.6 ohms. The temperature coefficient is .0053. What is its resistance at 0°? 40°? 80°?

847. Taking the specific resistance of copper as $17 \cdot 10^{-7}$, and temperature coefficient as $39 \cdot 10^{-4}$, and *assuming* this coefficient as *constant*, at what temperature would copper have no resistance?

848. The temperature coefficient of a certain iron wire is $53 \cdot 10^{-4}$. A coil of the wire has a resistance of 2000 ohms at 25°. What will be its resistance at 5°? 45°?

849. A coil of copper wire has a resistance of 2000 ohms at 16°. What is the range of temperature through which it may be used as a standard of resistance if the error must not exceed one-fourth of one per cent?

850. The temperature coefficient for a certain Cu wire is .0039; for a carbon filament it is —.0003. How many ohms of Cu resistance must be joined with a carbon filament of 100 ohms resistance so that the combined resistance may be constant?

851. Define the term *electrochemical equivalent*. State the relation between the electrochemical equivalent and the chemical equivalent.

852. The electrochemical equivalent of H is $1038 \cdot 10^{-8}$ (for 1 coulomb). The atomic weight of sodium is 23, its valence 1. Find the electrochemical equivalent of sodium.

853. A current of 2 amperes passes through a copper sulphate solution for 1 hour. If the anode is a copper wire, how much copper will be deposited on the cathode?

854. Compute the following electrochemical equivalents :

Substance.	Atomic Weight.	Valence.	Electrochemical Equivalent.
Hydrogen	I	I	$104 \cdot 10^{-7}$
Potassium	39.1	I	
Gold	196.2	3	
Copperic salts	63.18	2	
Copperous salts . . .	63.18	I	
Lead	206.4	2	

855. A deposit of 8.856 g. of copper is made by a current in $1\frac{1}{2}$ hours in a Cu–CuSO$_4$–Cu voltameter. What was the current strength?

856. A copper and a silver voltameter are placed in series. Find the ratio of the deposits formed.

857. Explain how you would arrange your apparatus in order to "plate" an article with silver.

858. A magnetic needle free to turn is placed in a uniform magnetic field. A new field at right angles to the first is then developed. Show by diagram what position the needle will assume. Does it depend on the pole strength or length of the needle? What would be the effect of reversing either field? both fields?

859. A wire carrying current is stretched north and south. The current flows from south to north. What position will a compass needle take when held *over* the wire? How will its position alter as it is brought nearer the wire? What position would it take if placed under the wire? if placed midway between two such wires carrying equal currents in the *same* direction? if in opposite directions? when between, but nearer to one than to the other?

860. A piece of wire 1 cm. in length is bent into a circular arc of 1 cm. radius. A current of 1 ampere flows in the conductor. What force would act on a + unit pole at the

center of the circle? What would be the field strength at the center of the circle when,

(a) $I = 1$ ampere, one complete turn?

(b) $I = 1$ ampere, n complete turns?

(c) $I = 1$ ampere, n turns, radius $= r$?

Note that 1 ampere $= \frac{1}{10}$ C.G.S. unit of current.

861. A circular coil of wire is placed in a north and south plane with its axis horizontal. A current is sent through the coil, flowing north on the upper side. What effect would the current have on a freely suspended magnetic needle when placed directly above the coil? directly below? in the same plane and just north? south? at the center?

862. What would be the strength of the magnetic field at the center of a coil of n turns, mean radius R, $I =$ one ampere? From this derive the law of a tangent galvanometer, consisting of one large coil and a short (?) needle at the center.

863. What do you mean by the term *constant of a galvanometer?* What is a tangent galvanometer? a sine galvanometer? Is a galvanometer of necessity one or the other?

864. Compute the current in each of the following cases, where $I_0 =$ galvanometer constant, $\delta =$ deflection in degrees:

Tangent galvanometer, $I_0 = 4.5$, $\delta = 25°$.

$I_0 = .035$, $\delta = 48°$.

$I_0 = 42.10^{-6}$, $\delta = 20°$.

What would the currents be if the galvanometer were a "sine" galvanometer?

865. When is a galvanometer said to be sensitive?

866. Explain how a sensitive galvanometer is constructed.

867. Explain how a given galvanometer may be made more sensitive.

868. If $I = 10 \dfrac{H}{\dfrac{2\pi n}{r}} \tan \delta$, $H = .19$, $n = 10$, $\delta = 25°$, what must be the radius of the coil if $I = 2$ amperes? How would δ be changed if H were reduced one-half?

869. A tangent galvanometer, $I_0 = 6 \cdot 10^{-3}$, $R = 200$ ohms, is placed in shunt with a resistance of 50 ohms. A deflection of 70° is observed. Find the *total* current.

870. A piece of soft iron is placed near a tangent galvanometer. What effect will it have on the galvanometer constant: (*a*) when placed in the same plane as the needle, and just north or south of it? (*b*) when in the same plane east or west? (*c*) when placed just below?

871. How would the action of the soft iron in Example 870 differ from that of a magnet?

872. The I_0 of a certain tangent galvanometer is $4 \cdot 10^{-3}$, where $H = .145$. What will I_0 be when the galvanometer is moved to a place where $H = .102$?

873. A current of .2 amperes causes a deflection of 40° in a tangent galvanometer where $H = .2$. What current would give the same deflection where H is .1?

874. The needle of a tangent galvanometer is observed to make 40 complete vibrations in one minute. I_0 at that point is $34 \cdot 10^{-6}$. When moved to another place it is found to make 25 complete vibrations in one minute. Find the constant in the new position.

875. (*a*) Give a diagram showing the construction of a simple type of tangent galvanometer. Explain in what position it must be placed in measuring current, and derive formula. (*b*) State the distinction between magnetic and diamagnetic substances. Describe an experiment by which the behavior of each, when placed in a magnetic field, can be shown.

876. A tangent galvanometer is connected in series with a generator of constant electromotive force and a known resistance which can be varied. A series of resistances are inserted, and corresponding deflections are observed. If these resistances are taken as x, and tangents of deflection as y, what sort of a curve will result? Does the entire curve have a physical meaning?

877. How is the *quantity* of electricity measured when it passes as an intense and variable current for a very short time. (Examples, condenser discharges and induced currents.)

878. What is a ballistic galvanometer? What is meant by the term *constant of a ballistic galvanometer?*

From the theory of the ballistic galvanometer we find that

$$Q = 10 \cdot \frac{T}{\pi} \cdot \frac{H}{G} \sin \tfrac{1}{2} \theta = Q_0 \sin \tfrac{1}{2} \theta,$$

while from the magnetic pendulum

$$T = 2\pi \sqrt{\frac{K_a}{MH}},$$

where H = horizontal component of earth's field.

T = periodic time of magnet in that field.

G = "true" constant of the galvanometer (tan.).

M = pole strength × distance between poles = ml.

K_a = moment of inertia.

θ = angle of maximum deflection.

We may write $\qquad Q_0 = 10 \dfrac{T}{\pi} \dfrac{H}{G}$ (1)

$$= \frac{T}{\pi} \cdot I_0 \qquad (2) \quad [I_0 = \text{tan. const.}$$

$$= \frac{10}{\pi} \cdot \frac{H}{G} \cdot \sqrt{\frac{K_a}{ml \cdot H}} \quad (3)$$

$$= \frac{10}{\pi} \frac{H^{\frac{1}{2}}}{G} \frac{K_a^{\frac{1}{2}}}{m^{\frac{1}{2}} l^{\frac{1}{2}}}.$$

879. Find the effect on Q_0 of increasing the horizontal intensity in any given ratio. Compare this with the change in I_0 due to the same increase in H.

880. The needle of a ballistic galvanometer is accidentally dropped; its pole strength is decreased. Will Q_0 be changed? I_0?

881. The constant of a ballistic galvanometer is .046 at a certain place. What will the constant be where H is nine times as great, if the needle is remagnetized and its magnetic moment increased fourfold?

882. The constant of a ballistic galvanometer at a point where T is 4 sec. is .045. What will the constant be where $T = 2$ sec.?

883. A coil of 100 turns, mean radius 40 cm., is turned 180° about a diameter which is perpendicular to the lines of force of a field of strength 10. The coil is connected with a ballistic galvanometer, and a deflection of 20° is observed. Resistance of the circuit 15 ohms. Find Q_0.

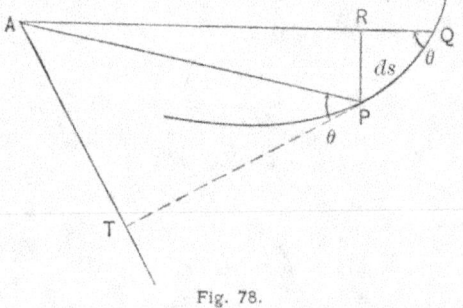

Fig. 78.

If ds is a current element so short that it may be regarded as straight, the laws concerning the magnetic force due to ds at any point A may be stated as follows:

(1) The force is $\perp$ to the plane APQ.

(2) The force is proportional to the length of ds.

(3) The force is inversely proportional to $\overline{AP}^2$.

(4) The force is proportional to the "broadside" projection of ds, i.e. to

$$PR = PQ \sin PQR = ds \sin \theta.$$

Summing up the last three of these four laws, we may say that

$$F_A = k \frac{ds \cdot \sin \theta}{r^2}.$$

[k depends on current strength and units used.]

The field at the point A is then found by integrating this expression. In order to perform the integration a relation between the variables must be given, *i.e.* the *shape* and *position*, with reference to A, of the circuit must be specified.

In the case of a very long straight wire, we have, if $p =$ perpendicular distance from A to the wire,

$$F_A = \int_{-\infty}^{\infty} kp \frac{ds}{(p^2 + s^2)} = \frac{2k}{p}.$$

For a wire of finite length $2\,l$, A in the plane perpendicular to its middle point, the limits would be $-l$ and $+l$.

884. Find how long the wire must be in order that when p is 5 cm. the field is within one per cent of that due to an infinite wire with the same current.

885. The horizontal component of the earth's field at a certain point in the Cornell Physical Laboratory is .145. At what distance from a long straight wire carrying 10 C.G.S. units of current would the field due to the current have the same intensity? (Here $k = 1$.)

886. What current must flow in an infinite straight wire that the magnetic field 10 cm. from the wire may exert a force on unit pole equal to the weight of 1 g. ?

887. Find the field strength at the center of a square, if a current passes around it.

888. Find the force exerted on a $+$ unit pole placed at the intersection of the diagonals of a rectangle, sides a and b, and carrying a current I.

889. Apply the formula $F_A = k \int \frac{ds \cdot \sin \theta}{r^2}$ to the case of a circular wire of radius R, when A is taken in the line perpendicular to the plane of the circle, and through its center. (Axis of coil.)

Show by diagram the direction of the force for each element, and for the complete circle.

What is the force *component along* the axis? Where is this a maximum?

MAGNETISM

890. State the law of attraction or repulsion between magnet poles. Where do similar laws occur in physics? Show how a definition of unit magnet poles follows directly from the law.

891. Find the force in dynes between two unlike magnet poles of strength 8 and 12 units respectively when the distance between them is .04 m.

The force varies according to the law $\frac{mm_1}{d^2}$.

Expressing d in centimeters $F = \frac{8 \cdot 12}{16} = 6$ dynes.

892. Two like magnet poles, of strengths 10 and 27 units respectively, are separated by a distance of 30 mm. Find the force in milligram's weight between them.

893. When two magnet poles are placed a distance apart of 1 cm. the force between them is 12 dynes. How must the distance be varied in order that the force may increase to 48 dynes?

894. What is a magnetic field of force? a magnetic line of force?

895. (*a*) Map the field of force around an ordinary bar magnet. (*b*) Map the field around two magnets placed with their like poles (supposed of equal strength) near each other and their axes at right angles.

896. A bar magnet is laid on a horizontal plane with its axis north and south, and its north-seeking pole north. Draw the resultant field, considering the earth's field as uniform.

897. In Example 896 find two points where the resultant magnetic force is o. Where would these points be if the magnet were reversed?

898. How does the distribution of lines of force due to a bar magnet differ from that of electric lines due to + and − induced charges on a cylindrical conductor?

899. A bar magnet is 40 cm. between the poles and pole strength 100, what is the direction and intensity of the magnetic force due to it at a point on the perpendicular to the line joining the poles and 50 cm. from this line?

900. Define *strength of field*. Find the force exerted on a pole of 12 units placed in a field of strength 326.

901. What is the strength of the magnet pole which is urged with a force of 2 mg. weight when placed in a field of strength .42?

902. What position does a magnet take when placed in a magnetic field (*a*) of which the lines of force are straight? (*b*) of which they are curved? Explain why the lines of force in a magnetic field can never cross.

903. Show that the number of lines of force coming from a pole of strength m is $4\pi m$.

904. The strength of a magnet pole is 72 units. Find the strength of field at a point 3 cm. away from it, assuming the other pole of the magnet to be so far away as to be of negligible effect at the point considered.

905. What are *consequent poles* in a magnet? How may they be produced?

906. How may a long magnet be placed with reference to a compass needle so that the needle is affected by one pole of the magnet only?

907. The angle of magnetic dip at Washington is 70° 18′, and the value of H is .2026. Find the total strength of field.

M

908. The angle of dip at New York is 70° 6′, and the total strength of field at that point is .61. Find the horizontal and vertical components.

909. Why is the earth's field simply directive in its action on a suspended magnet?

910. Why does not an ordinary compass needle dip or tend to dip?

911. Define *magnetic moment.* Find the dimensions of magnetic moment, and compute the moment of a magnet .13 m. long, and of pole strength 42, the magnetization being assumed uniform throughout the length of the magnet.

912. A magnet having a moment M is broken into n equal pieces of the same cross-section as the original magnet. What is the magnetic moment of each piece?

913. A magnet is placed in a uniform field of strength .362. When the axis of the magnet is normal to the direction of the field, the couple acting on the magnet is 2172 dyne-centimeters. Find the magnetic moment.

914. A magnet 10 cm. long has a pole strength of 60. When this magnet is placed in a field of strength .17, what is the couple acting on it (*a*) if the axis of the magnet be at right angles to the field? (*b*) if the axis be inclined at 45° to the field?

The force acting on each pole of the magnet is equal to the strength of the field × pole strength, *i.e.*,

$$F = Hm.$$

If the magnet lie at right angles to the field, this force is wholly effective in *turning* the magnet. If the magnet be inclined to the field by an angle θ the turning component of the force is less, being given by

$$F' = Hm \sin \theta,$$

and the moment of the effective couple is

$$F'l = Hml \sin \theta$$
$$= HM\theta$$

for small deflections.

The student should compare this result with the couple causing the vibration of an ordinary pendulum, and draw conclusions as to the character of the motion produced in each case. See 741, 742.

915. Show that the magnetic moment of a uniformly magnetized bar is proportional to the volume of the bar. Whence define *intensity of magnetization.*

916. A bar magnet has a cross-section of 1.2 sq. cm., a length of 12 cm., and a pole strength of 168. Assuming the magnetization to be uniform throughout the magnet, compute the intensity of magnetization.

917. Prove that the potential at a point distant r from a magnet pole of strength m is $\dfrac{m}{r}$.

918. In what units is magnetic potential measured? Find the potential of a point distant .6 m. from a magnet pole of strength 72.

919. Find the work done in carrying a pole of strength 4 units from a point distant 5 cm. from a magnet pole of strength 100 units to a point distant 2 cm. from this pole.

920. Find the potential at a point 6 cm. distant from the north pole of, and in line with the axis of, a bar magnet 10 cm. long and of pole strength 80.

921. A point P is distant OP from the center of a small magnet whose magnetic moment is M. Show that the potential at P is $\dfrac{M \cos \theta}{OP^2}$, where θ is the inclination of OP to the axis of the magnet.

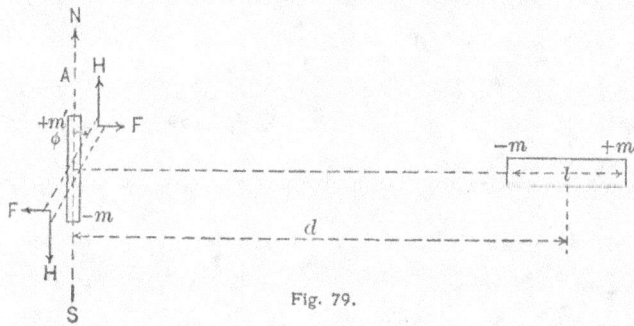

Fig. 79.

922. When the left hand magnet, Fig. 79, is so short compared with d that the lines joining their poles may be considered as parallel with that joining their centers, what is the torque exerted by the large magnet on the small one?

Treat force action of each pair of poles separately. Then take moments and add.

923. What torque is exerted by the earth's field?

924. By means of the last two examples show that when small magnet is in equilibrium

$$\frac{2\,ml}{H} = \frac{[d^2 - l^2]^2}{2\,d} \tan \phi.$$

[Where $l = \frac{1}{2}$ distance between poles of large magnet.

925. Explain why pole strength of small magnet need not be known. Why could it not be reduced to zero and yet have equation of Example 924 true?

926. Show that when d is very great in comparison with l,

$$\frac{2\,ml}{H} = \frac{d^3 \tan \phi}{2}.$$

927. If $H = .24$, $d = 1$ m., $l = 10$ cm., $\phi = 25°$.

What is the pole strength of the magnet?

928. Prove that when magnets are placed as in Fig. 80 [the length of the small magnet being small compared with d]

$$\frac{2\,ml}{H} = [d^2 + l^2]^{\frac{3}{2}} \tan \phi.$$

929. When l may be neglected, show that

$$\frac{2\,ml}{H} = d^3 \tan \phi.$$

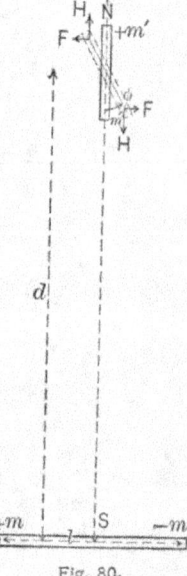

Fig. 80.

930. How do the results of Examples 926 and 929 compare. Explain why such a difference should be expected.

931. If the large magnet were reversed, what change of position would the small one experience?

932. If the magnets were exactly alike, and each were sus-pended so as to be free to move, would each turn through the same angle in Fig. 79? in Fig. 80?

933. Taking axes parallel and normal to the axis of a magnet, plot curves showing (*a*) the variation of potential and (*b*) the variation of magnetic force with distance along the axis. Dis-cuss the relation existing between these curves. (Only one pole of the magnet is to be considered.)

934. Define magnetic induction (*B*), permeability (*μ*), and susceptibility (*κ*). Imagine a piece of soft iron placed in a weak field. Further, imagine the field to gradually increase in strength. Show by means of a curve the changes which take place in the induction in the iron with the increase in the field strength.

Such a curve is called a **magnetization curve,** and is of great practical value. It is usually plotted with the induction *B* and the field strength *H* as co-ordinates. Obviously the ratio of any ordinate *B* to the corresponding abscissa *H* is the permeability *μ* of the iron.

935. Which is the more easily magnetized, soft iron or steel? Which retains the greater amount of magnetism when the mag-netizing force is removed? Explain answers fully in accordance with the molecular theory of magnetism.

936. Why is magnetism removed by heating? Why are iron rods subjected to tapping or jarring liable to become magnetized?

937. An iron tube is driven into the earth in the Northern Hemisphere. What would be its magnetic condition?

938. What kind of iron would you choose for the construction of permanent magnets? of telegraph instruments?

939. Show that *B*, *H*, and *I* are quantities of the same kind or dimensions. What must therefore be true of *μ* and *κ*?

940. Explain the principle of magnetic screening, as when a galvanometer needle is protected by an iron screen.

941. A sample of transformer iron gives the following data. Plot and discuss the magnetization curve.

H	B
1.32	1324
2.0	3650
4.64	8800
7.1	10980
10.73	12450
14.65	13320
19.42	13920
37.0	15032
49.8	15465

942. Compute the data requisite to plot a permeability curve, using H and μ as co-ordinates.

943. Discuss the equation $B = H + 4\pi I$, explaining the meaning of each term.

944. A sample of iron shows $I = 1226$ for $H = 40$. Compute the susceptibility; the induction; the permeability.

945. Show that the force with which a magnet attracts its keeper is

$$F = \frac{B^2}{8\pi} A,$$

stating clearly the conditions that must be fulfilled in order that this equation may be true.

946. It is found that when the poles of a certain magnet are reduced in area the lifting power of the magnet is increased. Why is this?

947. A certain magnet having a pole face of area 4 sq. cm. is found to sustain a maximum load of 2 kg. Find the induction.

948. What is meant by the term *magnetomotive force?* What is the magnetic analogue of Ohm's law?

949. The field magnet of a dynamo is wound with 3200 turns of wire. The normal field current is 820 milliamperes. What is the number of ampere turns?

950. A circular ring of iron has a cross-section of 8 sq. cm. and a mean radius of 7.5 cm. What magnetomotive force must be used to set up a total magnetic flux of 120000 lines? The permeability for this induction is 526.

951. If an air gap is cut in a magnetic circuit, how is the magnetization curve affected?

952. A current flowing in the turns of a short solenoid produces a field of a given strength along the axis. When an iron core is inserted, the value of H is changed. Why is this?

953. A certain magnetic circuit has a cross-section of 36 sq. in. It is made of cast iron, showing a permeability of 71 for a magnetizing force of 127. Compute the total magnetic flux (or induction).

954. A long solenoid is wound with 20 turns per cm. Compute the value of H along the middle of the solenoid, (a) when no iron is present, (b) when iron giving the data of Problem 941 is present, the current in both cases being 5 amperes.

955. What is *hysteresis?* What is represented by the area of a hysteresis loop?

956. A transformer core contains 3840 cu. cm. of iron. The hysteresis loss is 16300 ergs per cycle per cubic centimeter. If this transformer be supplied with an alternating current of frequency 120 periods per second, what is the power (in watts) lost in hysteresis?

957. How does the energy spent in hysteresis appear? What is the effect of jarring on hysteresis?

958. State clearly the meaning of the symbols in the formula for the magnetic pendulum, $T = 2\pi\sqrt{\dfrac{K_a}{MH}}$.

959. Explain how the magnetic pendulum differs from the gravitational pendulum. Would there be any objection to using a magnetic pendulum for a clock?

960. What must be the pole strength of a magnet, moment of inertia 1800, distance between the poles 10 cm., that it may make 20 complete vibrations in 4 min., where $H = .145$?

961. A large block of soft iron is placed beneath a horizontal vibrating magnet. What will be the effect on T?

962. A magnet is set in vibration where H is .16, and T is found to be 3 sec. When taken to another place, T' is found to be 3.2 sec. Find H'.

963. Derive the equation $T = 2\pi\sqrt{\dfrac{K_a}{MH}}$, explaining any approximations or assumptions made.

964. If a magnet is struck several blows, what will be the probable effect on its time of vibration as a magnetic pendulum?

965. A strip of lead is bound to a magnetic pendulum. What is the effect on T?

In the study of the magnetic forces due to currents, of tendencies of conductors carrying current to move in a magnetic field, and of the direction of induced currents, it will be found that the concept of lines of force is one of great utility. Remembering that two magnets placed parallel, with their *like* poles contiguous, will tend to separate, we see that if this action is to be ascribed to a property of lines of magnetic force we should say that lines of force parallel and in the same direction repel.

It will be found convenient to suppose that the characteristics of lines of magnetic force are in part as follows:

(a) Magnetic lines of force parallel and in *same* direction *repel* each other.

(b) Magnetic lines of force parallel and in *opposite* directions *attract*.

(c) Magnetic lines of force are similar to *tense*, elastic threads which first bend, and then break when a conductor moves across them.

(d) These lines tend to shorten and also to pass through iron rather than air.

These, together with the fact that when a current flows lines of magnetic force tend to form

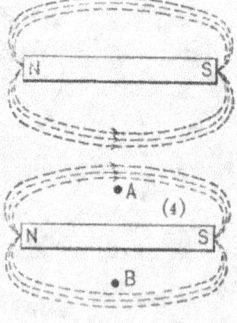

Fig. 81.

circles around it, are very useful in indicating the relations of currents to varying fields, etc.

The direction of current and the positive direction of the lines of force due to it are related to each other in the same way as are the direction of translation of a *right-handed* screw, and the direction in which it is turned. Or, if

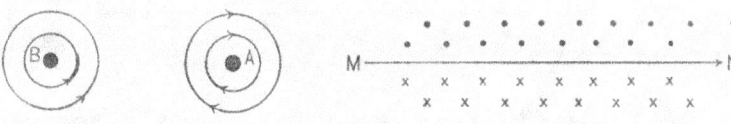

Fig. 82.

we imagine current to flow from the eye to a clock-face, lines of force around the current would be such that a + pole would go around it like the hands of the clock or "clockwise." If current pass down perpendicular to the paper at *A*, the entire plane has lines directed as shown.

For convenience in diagram, we shall indicate that a *line of force* is coming *up* through the paper by a •, going down by a ×. Thus, if current flows in the line *MN* in the plane of the page, the magnetic lines are vertical circles encircling *MN* clockwise, looking from *M* to *N*.

This is not suggested as the only way in which these relations may be remembered, but as one found of considerable convenience in practice.

A few diagrams are added to show the application of these statements.

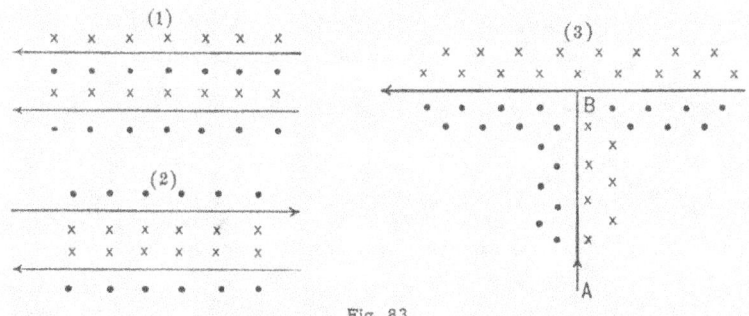

Fig. 83.

(1) Two parallel *currents* in the same direction *attract*.

(2) Two parallel *currents* in opposite directions *repel*. Likewise for conductors inclined to each other.

(3) Two rectilinear currents perpendicular to each other. *AB* free to turn about *A*. *B* moves to the left. Similarly, if *CD* is a circle and *AB* a radial current.

(4) Current down perpendicular to plane of magnet. At *A* conductor and magnet tend to approach; at *B* to separate. (See Fig. 81.)

The property of magnetic lines of force assumed in (*c*) may be conveniently used in determining the direction of induced currents. We might look at

the matter of relative motion of a conductor and lines of magnetic force somewhat, as indicated by Fig. 84.

Let A be the intersection of a conductor with the plane of the paper, and
let the lines of force be parallel to this plane. When A is moving to the
right or the field moves to the left, we may consider the lines of force from

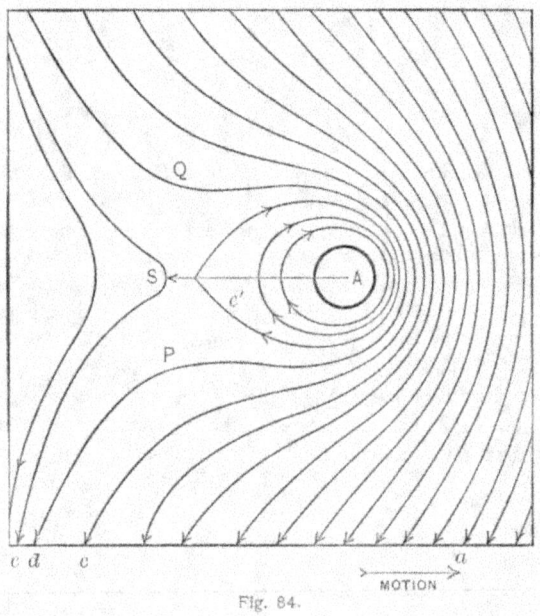

MOTION

Fig. 84.

a to c as crowded together and stretched. d is stretched so far that lateral
compression is forcing it to encircle A, e has gone through the phases b, c, d,
and the points corresponding to P and Q of d have united as at s, leaving e'
encircling the wire. Current tends then to flow down, just as current would
flow to set up like lines or to
oppose the motion.

966. The case of an
east and west wire in
the earth's field is a
good example. If MN
and OP (Fig. 85) repre-
sent two lines of the
earth's field, AB an east
and west wire, then, if

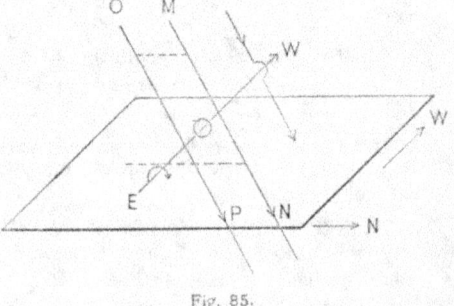

Fig. 85.

AB is moved up, the lines tend to encircle it as shown. Which way does current tend to flow? Does the current help or oppose the motion?

967. Draw the diagram when the wire is falling.

968. A telegraph wire is stretched east and west. The direction of the earth's field is 75° with the horizontal. Show by diagram the direction of the induced currents

        (*a*) When it falls vertically downward.

        (*b*) When it is raised vertically.

Show also in what direction to move it in order to get a maximum current; a minimum current.

969. Two parallel wires are placed as in Fig. 86. When the key k is closed, what takes place in the other wire? If the wires moved apart with a velocity equal to that of light, would the same effect be observed?

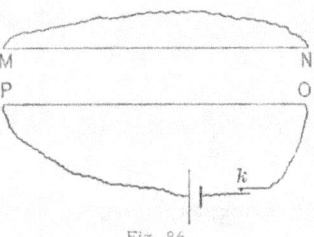

We may consider circular lines of magnetic force as springing out from the first wire. Their radii increasing at what rate?

Fig. 86.

970. The north pole of a magnet passes through the bottom of a cup C. Mercury covers the bottom, and a wire suspended

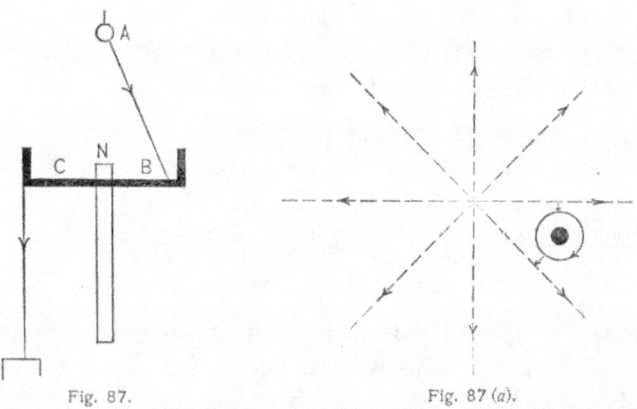

Fig. 87.   Fig. 87 (*a*).

vertically above N dips below the surface of the mercury. If current flows from A to B, show that B will move away from and rotate around N.

Consider the projection of the lines of force due to the magnet on the surface of the mercury. (See Fig. 87.)

971. Extend to the case of a flexible conductor.

The student should apply this method to cases of action of magnetic fields described in text-books or observed in lectures.

972. A solenoid is placed with its axis north and south; its terminals are connected with a galvanometer. When a piece of soft iron is thrust into or drawn from the coil, an induced current is observed. Explain. Would the effect be increased or diminished if the axis of the solenoid were east and west?

973. A small piece of soft iron is suspended near a magnet by a thread. Explain the position it will take by reference to (d).

974. Explain why a solenoid tends to shorten when current is passed through it.

975. Explain the effect of a copper box surrounding a vibrating magnetic needle.

976. A metal plate is revolved between the pole piece of an electromagnet. It is observed that it is harder to maintain its motion when current is passing through the coils of the magnet. Explain this. What becomes of the energy used in turning the plate? Does the magnet tend to move?

977. Show in what direction a magnet may move with reference to a fixed wire in order that no electromotive force may be set up in the wire.

978. In the figure of Example 1000, in what direction must the coil turn that current may flow from A to D?

979. A solenoid is wound so that it looks like a right-handed screw. An iron core is placed in it and you are required to make a given end a north-seeking pole. Give a diagram showing the direction of the current.

980. Two points of different electrical potential are joined by

(a) a straight wire,

(b) a coil of wire,

(c) a coil of wire with a soft iron core,

(d) a coil of wire with a permanent magnet as a core.

Indicate the differences in the magnetic fields produced in these cases.

981. (a) A wire perpendicular to the plane of the paper carries current downward. Indicate form and direction of the lines of magnetic force. (b) A parallel wire carrying current in the same direction is brought near. How is the field altered? What action takes place between the wires?

982. (a) Define *permeability*. (b) Draw the lines of force for the magnetic fields shown in diagrams, Fig. 88. (c) What is the power of energy in the case of an induced current produced by motion in a magnetic field? (Winter, '96.)

983. Find the force acting on a pole of 60 units' strength at a distance of

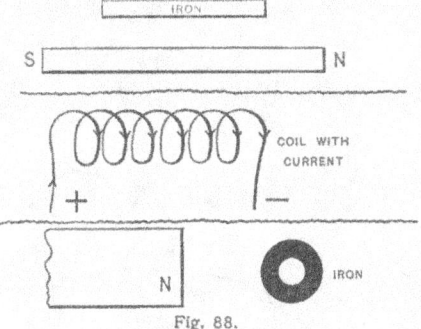

Fig. 88.

5 cm. from an infinitely long straight conductor carrying a current of 5 amperes.

984. To reduce the force in the foregoing case by one-half, where must the pole be moved?

985. A bar magnet is allowed to drop vertically through a closed loop of wire. What are the directions of the induced currents?

986. A certain wire is moved through a magnetic field so as to cut 10^9 magnetic lines of force in 2 sec. What is the average electromotive force induced?

The E.M.F. induced is proportional to the *rate of cutting*. To reduce the result to practical units (volts), divide by 10^8.

987. A wire 30 cm. long is moved through a field of strength 6000 lines per sq. cm. at the rate of 10 m. per second. Find the induced electromotive force in volts.

988. A centimeter length of a straight wire is placed at right angles to the lines of force of a uniform magnetic field. 1 C.G.S. unit of current flows through the wire. The strength of the magnetic field is 1000. What force acts on the wire? If the current is ten times as great, the field one-tenth as strong, and the wire 1 m. long, what force would act?

989. If a wire 1 m. long, current of 100 amperes, is placed horizontally at an angle of 30° with a uniform horizontal field, what force acts on the wire if the field strength is 1000? In what direction does it act?

990. A flat loop of wire of resistance .001 ohm, and area 1 sq. m., rests on a horizontal table. If the loop be picked up and turned over, what is the total quantity of electricity set in motion?

991. Would it make any difference in the quantity if the loop were turned slowly or quickly?

992. How can a straight wire be moved in a magnetic field, and yet have no electromotive force developed in it?

993. If a closed loop of wire be moved without change of plane through a magnetic field of uniform strength, will any current flow in it? Will any electromotive force be developed in it?

994. A wire 2 m. long, and lying horizontally east and west, is allowed to fall freely. (*a*) Find the value of the induced electromotive force at the end of 3 sec. (*b*) Find the mean value of the induced electromotive force during a fall of 5 sec. (*c*) Find the time elapsing before the electromotive force shall be just 1 volt.

995. *AA'* and *BB'* are a pair of copper rails, so large that their resistance may be neglected in comparison with that of the rest

of the circuit. S is a wire of resistance 1 ohm, sliding without
friction over the rails, and at right angles to them.   Resistance
of galvanometer circuit, 3 ohms.   If the rails are in a field of
3000 lines per sq. cm., the direction of the field being upward,

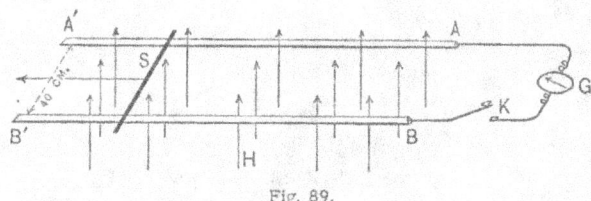

Fig. 89.

normal to the plane of the rails, and the distance between the
rails be 40 cm., find :

(*a*) The velocity required to develop an electromotive force
in S of 1 volt.

(*b*) The direction of this electromotive force when the motion
is in the direction indicated.

(*c*) The current in the circuit when k is closed.

(*d*) The work done in the circuit.

(*e*) The force necessary to propel S at this velocity.

996. Show that the quantity of electricity set in motion by
any displacement of the slider is independent of the velocity
with which that displacement takes place.

997. If the velocity of the slider were doubled, what would
be true of the work done in the circuit ?

998. If the galvanometer were replaced by a cell developing an
electromotive force of 1 volt, and having a resistance of 3 ohms,
in what direction and with what velocity would the slider move ?

999. How can the slider and rails of Problem 995 be used to
show that the dimensions of resistance in the electromagnetic
system are those of a velocity ?

1000. A rectangular loop of wire .1 m. wide and .2 m. long
rotates uniformly at a speed of 1200 revolutions per minute in a

field of 4000 lines per square centimeter.   Find the average
value of the electromotive force
induced.

Since all that is desired is the *average* value of the induced electromotive
force, we have only to find the total
change in the number of lines thread-
ing the loop per revolution, and divide
this by the time of one revolution.

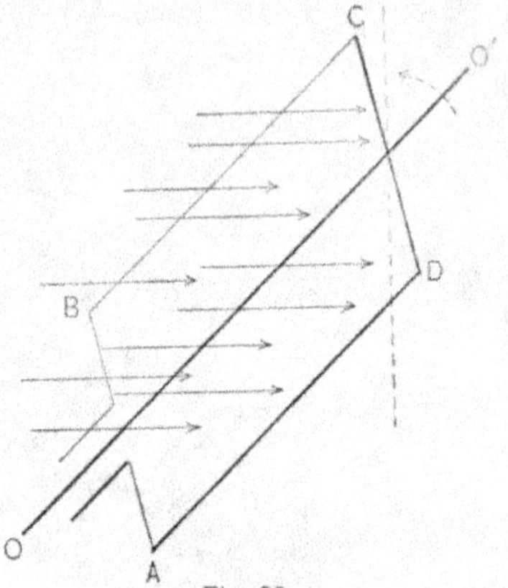

1001. With the direction of
field and of rotation as indicated,
what is the direction of the induced electromotive force?

1002. When such a coil rotates in a uniform field, to what

Fig. 90.

are the instantaneous values of electromotive force proportional?

1003. If a loop of wire rotating in a magnetic field form part
of a closed circuit, the resulting current is an alternating one.
Sketch and describe a device by which the current may be
caused to flow always in the same direction in the external
circuit.

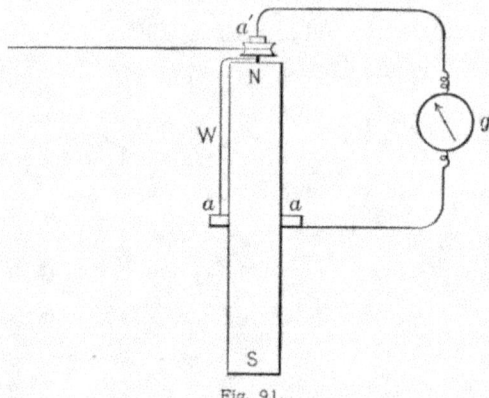

Fig. 91.

1004. A wire *w* is caused to rotate around the north pole of
a magnet by means of a cord on a pulley.   Contact is made in

the mercury cups a, a', the closed circuit being $aa'g$. The strength of pole is 72. The wire is caused to rotate with a speed of 600 revolutions per minute. The resistance of the circuit is .01 ohm. What is the current in amperes?

Would current flow if the wire extended the entire length of the magnet?

1005. If the wire were fixed and the magnet were placed on a pivot so as to be free to turn about its axis, what would happen when current is passed through the wire?

1006. A Faraday disc has a radius of 15 cm. It rotates with a speed of 2400 revolutions per minute in a field normal to the disc of average density 2000 lines per square centimeter. Compute the electromotive force of the machine.

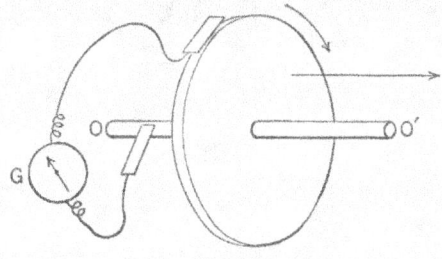

Fig. 92.

1007. What essential differences are found in the following types of dynamos : (*a*) magneto, (*b*) series, (*c*) shunt, (*d*) compound?

1008. What type of dynamo is best adapted to incandescent lighting?

1009. Which would suffer most from a short circuit, a shunt or a series dynamo?

1010. What is meant by residual magnetism? What important part does it play in the operation of dynamos?

1011. A certain series-wound dynamo refuses to generate. The connections of the field coils are reversed, when the machine immediately "picks up." Explain. Would reversing the direction of rotation have the same effect?

N

1012. A bipolar dynamo has upon the surface of its armature 480 conductors; and the armature rotates with a speed of 1200 revolutions per minute in a total magnetic flux of 1250000 lines. Compute the electromotive force of the machine.

1013. What difference exists between the ring (Gramme) and drum armature windings?

1014. A ring armature of 320 turns rotates with a speed of 1800, while a drum armature of 240 turns rotates with a speed of 1200. The field being the same for both armatures, compare the E.M.F. developed.

1015. Arc lights are usually run in series. Does the armature of an arc-lighting dynamo need to be wound with fine or coarse wire? Is a high degree of insulation necessary? Are few or many turns of wire required?

1016. Glow lamps are run in parallel. Answer the questions of the last problem, with reference to a dynamo for incandescent lighting.

1017. In what three ways may the electromotive force of a dynamo be increased?

1018. What fixes the maximum current output of a dynamo?

1019. What should be the characteristic features of a dynamo designed for electric welding?

1020. The field circuit of a dynamo has the form shown in Fig. 93. It is required to find the number of ampere turns needed on the field limbs to set up in the air gap a magnetic density of 6000 lines per square centimeter. Concerning this machine the following data are known:

Diameter of armature core . . . .   25 cm.
Length of armature core . . . .   36 cm.
Mean length of magnetic circuit in field (*i.e.*
    dotted line *abcd*) . . . .   145 cm.

Permeability of armature iron for a magnetic
    density of 6000 1120
Coefficient of magnetic leakage for this type of
    circuit 1.5
Permeability of field iron for a magnetic density
    of 1.5×6000 2250
Depth of double air gap . . . . 0.72 cm.

The work done in carrying a + unit magnet pole around the path indicated
by the dotted line is

$$\frac{4\pi Si}{10} = \int H dl,$$

where S is the number of turns of wire on the
field, and i the current in them. Considering
the magnetic circuit as made up of three separate parts, in each of which the value of H is
assumed to be constant, we have

$$\frac{4\pi Si}{10} = H_a l_a + H_g l_g + H_f l_f,$$

the subscripts a, g, and f referring to the armature, air gap, and field, respectively.

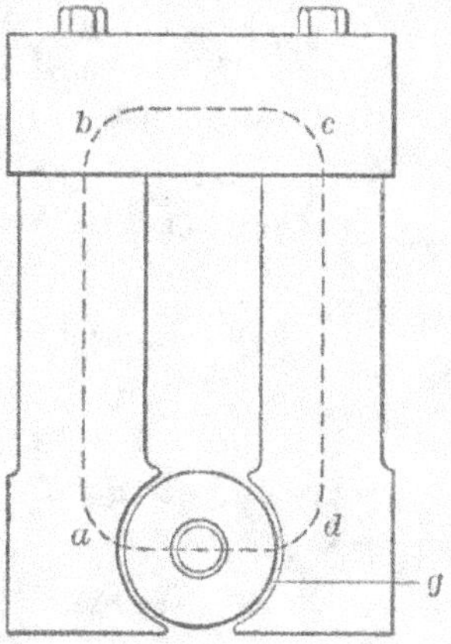

Taking the computations in the order indicated, we have

Fig. 93.

$$H_a = \frac{B_a}{\mu_a} = \frac{6000}{1120},$$

and

$$H_a l_a = \frac{6000}{1120} \cdot 25 = 134 -.$$

For air, $\mu = 1$,

hence $H_g l_g = 6000 \times 0.72 = 4320.$

Now in every dynamo there is a certain amount of stray field, or waste
magnetic flux, which forms closed loops by various paths outside the air gap.
The amount of stray field is readily found for different types of machines by
experiment. The ratio $\dfrac{\text{total magnetic flux}}{\text{useful magnetic flux}}$ is called the coefficient of magnetic leakage. The induction to be provided for in the field is, therefore,

$$k B_a = 1.5 \times 6000 = 9000,$$

and we have $$H_f l_f = \frac{9000}{2250} \cdot 145 = 580,$$

$$\Sigma H l = 134 + 4320 + 580$$

$$= 5034.$$

The requisite number of ampere turns is therefore

$$Si = \frac{5034}{1.26} = 4000 \text{ nearly.}$$

The student should note that in the foregoing method certain assumptions are made which are not rigorously true. The method, however, gives results which meet all the requirements of practical dynamo design.

1021. The armature of this dynamo has upon its surface 184 conductors, and it makes 1200 revolutions per minute. Compute the electromotive force.

Since the pole pieces are not likely to cover more than 80 per cent of the armature, the magnetic density may be taken, as in the preceding case, as the same in air gap and armature. The cross-section of the armature is

$$25 \times 36 = 900 \text{ sq. cm.}$$

The total number of lines is therefore

$$900 \times 6000 = 54 \times 10^5.$$

The total electromotive force developed is

$$\frac{NCn}{10^8},$$

where N is the total flux, C the number of conductors on the armature, and n the number of revolutions per second. This gives

$$\frac{54 \times 10^5 \times 184 \times 20}{10^8} = 200 \text{ volts, nearly.}$$

1022. It is found that over and above friction a certain amount of power is required to turn the armature of a dynamo when the machine is on open circuit. To what two causes is this waste of power due? How may it be diminished?

1023. What is meant by a *characteristic curve?* A series machine gives the following data. Plot it, using current on the X-axis.

Potential Difference.	Current.
2.6	0
10.3	4
31.4	10
43.5	14
52.3	20
56.1	25
60	34
62	45

1024. This machine would work unsatisfactorily below 40 volts. Why?

1025. Suppose a line to be drawn from any point on the characteristic to the origin. What is indicated by its pitch?

1026. The product of the co-ordinates of any point on the curve is taken. What is shown by this product?

1027. The data in the first column are potential differences at the terminals. Given that the internal resistance of the machine is .2 ohm, how may the total electromotive force be found?

1028. When the circuit of a series machine is closed through a given resistance, why do not the current and electromotive force continue to increase indefinitely?

1029. What is the general shape of a shunt characteristic? What would be the characteristic of a perfectly "compounded" dynamo?

1030. What is meant by the gross efficiency of a dynamo? the net efficiency? the electrical efficiency?

These terms are defined by the ratios:

$$\text{Gross efficiency} = \frac{\text{total electrical energy developed}}{\text{total mechanical energy supplied}}.$$

$$\text{Net efficiency} = \frac{\text{useful electrical energy developed}}{\text{total mechanical energy supplied}}.$$

$$\text{Electrical efficiency} = \frac{\text{useful electrical energy}}{\text{total electrical energy}}.$$

Since every machine has some internal resistance, the electrical efficiency can never reach 100 per cent.

1031. A certain dynamo develops electric power to the amount of 10 kilowatts. If the gross efficiency of the machine is 85 per cent, how many horse-power must be furnished to drive it?

1032. The internal resistance of a series dynamo is .2 ohm. The machine develops a maximum current of 40 amperes at an *available* potential difference of 100 volts. What is the electrical efficiency?

1033. The net efficiency of a certain dynamo is 70 per cent; the gross efficiency is 84 per cent. What is the electrical efficiency of the machine?

1034. A certain dynamo requires 8 kilowatts when driven at full capacity. The net efficiency being 82 per cent under these conditions, and the pressure at the terminals being 105 volts, what is the maximum current output?

1035. A shunt dynamo has a field resistance of 70 ohms, and an armature resistance of .022 ohm. When running at full load the machine develops 80 amperes at an available potential difference of 110 volts. What is the electrical efficiency of the machine?

1036. A house is to be lighted with 40 glow lamps, each requiring .5 ampere and 110 volts. Allowing for a loss of 4 per cent in the mains, a net efficiency in the dynamo of 84 per cent, and a reserve power in the engine of 15 per cent more than that actually required to run the lamps, what should be the horse-power of the engine installed?

1037. What determines the practical limit of long-distance transmission of power?

1038. When current is supplied to a direct-current dynamo it runs as a motor. Explain by reference to Problem 995.

1039. An ammeter is introduced into a motor circuit. The current is found to be stronger when the armature is held still than when it is allowed to run. Explain.

1040. If the wheels of a street car were securely locked, the controller could not safely be turned so as to let maximum current flow. Why?

1041. A wire 1 m. long, carrying a current of 20 amperes, is held in a uniform field of 6000 lines per square centimeter. Find the restraining force.

To obtain the force in dynes, the current must be reduced to C.G.S. units, *i.e.* must be divided by 10.

1042. If the field of a motor be strengthened, will it run faster or slower, other conditions remaining unaltered?

1043. Assuming that the energy absorbed by a motor appears in two ways only, namely, as useful work and as heat due to resistance, show that the motor does maximum work when the counter electromotive force is one-half the impressed electromotive force.

Let E be the constant impressed electromotive force, i the current, r the internal resistance of the motor, and e the counter electromotive force. We have, according to the foregoing assumption, total power absorbed $= Ei = ei + i^2r$, whence useful power $= w = Ei - i^2r$. i being the only variable in the right-hand number, we have merely to find the value of i to give maximum w.

1044. Show that it follows from the foregoing that the efficiency of a motor doing maximum useful work is but 50 per cent.

1045. Under what conditions will a motor run at maximum efficiency?

1046. A series-wound motor has a resistance of .2 ohm. When supplied with 5 amperes at a potential difference of 110 volts, what is the energy wasted in heating? Of the energy not wasted in heating 92 per cent is used in overcoming the torque due to friction hysteresis and eddy currents. What is the net efficiency of the motor?

1047. A motor is supplied with a current of 15 amperes at a pressure of 110 volts. The power developed at the pulley is 1.81 horse-power. Compute the net efficiency of the motor.

1048. If two armatures were mounted on the same shaft, would it be possible to use one as motor and the other as a dynamo? What would such an arrangement be called, and what uses might it have?

1049. (a) What is meant by the period of an alternating current? (b) A small 8-pole alternator makes 1800 revolutions per minute. What is the periodicity of the current developed?

(b) Eight poles, alternately north and south, give 4 complete periods per revolution; hence the periodicity, or frequency,

$$= \frac{4 \times 1800}{60} = 120.$$

1050. Find the mean value of an harmonic or sine electro-motive force.

Instantaneous values being given by

$$E = e \sin \alpha,$$

we should have as the mean e

$$\frac{E \int_0^\pi \sin \alpha \, d\alpha}{\int_0^\pi d\alpha},$$

which is readily integrated.

The mean value of an harmonic current is similarly found from the expression

$$i = I \sin \alpha.$$

NOTE. — In the treatment of alternating currents it is usually justifiable to consider them as harmonic even though they depart somewhat from the sine law. In the following problems the current is assumed to be a simple sine function of the time.

1051. The maximum value of an alternating current is 120 amperes. What is the mean value?

1052. What is the maximum value of an alternating current that will cause the same quantity to flow across any cross-section of a conductor in a given time as does a direct current of 63.6 amperes?

1053. An alternating current has a maximum value of I. What is the value of the direct current that will develop the same heat in any given resistance?

By Joule's law the heat developed is proportional jointly to the square of the current and to the resistance of the circuit. If the current be a varying one, the heat is proportional to the *mean square*. We therefore have to find the value of

$$\frac{I^2 \int_0^\pi \sin^2 \alpha \, d\alpha}{\int_0^\pi d\alpha},$$

which is the mean square of a current whose maximum value is I.

The "square root of the mean square" of an alternating current is called its *virtual* value, and is of great importance.

1054. The *virtual* value of an alternating current is 35.3 amperes. What is its maximum value? its mean value?

1055. Which will develop the greater amount of heat in a given circuit, a direct current of 50 amperes, or an alternating current whose *mean* value is 50 amperes?

1056. What is meant by self-induction? Give two definitions of the *coefficient of self-induction*. Define the *henry*.

1057. The field magnet of a shunt dynamo consists of an iron core wrapped with a great many turns of fine wire. If a current be sent through such a field for an instant by striking the proper wires across one another, only a slight spark is observed; but if the current be allowed to flow for a second and then the circuit be broken, a heavy spark is obtained. Explain.

1058. If a current of 2.1 amperes flowing in a coil of 100 turns set up through that coil a magnetic flux of $.084 \times 10^8$ lines, what is the coefficient of self-induction of the coil, assuming the coil to contain no iron?

If the circuit were broken, the wire composing it would be cut by $100 \times .084 \times 10^8$ lines. The change in the current is 2.1 amperes. Therefore the inductance of the circuit is

$$\frac{100 \times .084 \times 10^8}{2.1 \times 10^8} = 4 \text{ henrys.}$$

1059. An harmonic current of 20 amperes (virtual value) is flowing in a given circuit. If the frequency be 120 periods per second and $L = .06$ henry, what is the electromotive force of self-induction?

1060. If the resistance of the foregoing circuit be 2.4 ohms, what is the value of the electromotive force impressed on the circuit?

1061. Find the impedance of a coil having a resistance of 40 ohms and an inductance of .6 henry. Frequency of the alternating current 120.

1062. The resistance of a given coil is 8 ohms, inductance, .3 henry. Compute the angle of lag for an alternating current of frequency 100.

1063. The current in a coil is 40 amperes; the potential difference around the terminals of the coil is 102 volts. The angle of lag is found to be 36°. Compute the power.

1064. Show by a diagram what is meant by the *lagging* of an alternating current behind the impressed electromotive force.

1065. To obtain the power spent in a circuit in which a direct current of constant value is flowing, it suffices to take the product ei. Explain why this is usually incorrect in the case of an alternating current.

1066. An alternating current of frequency 120 periods per second is passing through a straight wire of negligible inductance. When the wire is coiled around an iron core, the current is observed to fall off 40 per cent. The resistance of the wire being 6 ohms, what is the inductance of the coil?

1067. What are the essential features of a transformer, and what advantages arise from its use?

1068. In what four ways is energy wasted in a transformer?

1069. The ratio of the primary and secondary turns of a transformer is 20 : 1. If at full load, the primary power is 4000 watts and the primary current 2 amperes. What are the values of the secondary E.M.F. and current, the efficiency of the transformer being 90 per cent?

1070. What is necessary that an ordinary alternator may run as a motor?

1071. What is meant by a rotating magnetic field? How may it be produced?

1072. A magnetic field whose instantaneous strength is given by the equation

$$b = 6000 \sin wt$$

is combined at right angles with another of strength

$$b' = 5000 \sin \left(wt - \frac{\pi}{2} \right).$$

Find the magnitude of the resultant field.

1073. What are the important differences between synchronous motors and induction motors?

Magnetic and Electrical Units. — We have seen how from the arbitrarily chosen units of mass, length, and time a convenient and consistent system of mechanical units is built up. From the same fundamentals, and in a similar way, the units necessary for electrical and magnetic measurements may be derived. In every case the definition of the unit is based on a physical law or a deduction from a physical law. It is evident that more than one unit might easily be chosen according as different physical phenomena were made the basis of the selection. Thus two distinct C.G.S. systems of electrical units have arisen. One, the *electrostatic* system, is based on the definition of unit quantity of electrification as defined from the experimentally proved relation between the magnitudes of electric charges and the force, *in air*, between them. This relation is

$$f \propto \frac{qq'}{l^2}.$$

Now since unit length is a fundamental, and unit force has been already chosen, it is consistent to say that unit quantity is such a quantity that acting on an equal quantity at unit distance will repel it with a force of one dyne. Unit quantity is thus made to depend directly upon the units of force and distance. To ascertain the way in which the fundamentals are involved in any measurements of quantity we must pass to dimensions; thus,

$$\text{unit force} = MLT^{-2} = \frac{Q^2}{L^2}$$

whence $Q = M^{\frac{1}{2}}L^{\frac{3}{2}}T^{-1}$.

Unit current is said to flow in a circuit when unit quantity is conveyed in unit time. This makes the dimensions of current

$$\frac{Q}{T} = M^{\frac{1}{2}}L^{\frac{3}{2}}T^{-2}.$$

PROBLEM. — Suppose that the unit of time were increased threefold, and the unit of length were doubled. How would the C.G.S. electrostatic unit of current be affected?

Making these changes in the fundamentals, we have for the new unit of current

$$M^{\frac{1}{2}}(2\,L)^{\frac{3}{2}}(3\,T)^{-2} = .314\,M^{\frac{1}{2}}L^{\frac{3}{2}}T^{-2}.$$

That is, the new unit is smaller than the old, the ratio being $\frac{314}{1000}$.

Hence a given current would appear to be $\frac{1000}{314}$ times as great.

The other system is called the C.G.S. *electro-magnetic* system. The primary definition is that of unit current, based on the action between an electric current and a magnet-pole in its vicinity. It is known, as the result of experiment, that a magnet-pole placed at the center of a loop of wire carrying current is urged along the axis of the loop, *i.e.* at right angles to the plane of the loop, with a force which varies as the current, the strength of the magnet-pole, and the length of the wire directly, and as the square of the radius of the loop inversely. That is,

$$F = K\frac{I\,2\,\pi r m}{r^2},$$

$$I = K'\frac{Fr}{2\,\pi m}.$$

If I be such that when m is a unit, magnet-pole and r is unity, the force is $2\,\pi$ dynes, then

$$I = K'.$$

And if it be agreed to call this current unit current, then any current thereafter is given by

$$I = \frac{Fr}{2\,\pi m}.$$

The dimensions of unit current are

$$\frac{\text{force} \times \text{distance}}{\text{strength of pole}},$$

or $$M^{\frac{1}{2}}L^{\frac{1}{2}}T^{-1}.$$

The quantity conveyed by unit current in unit time is taken as unit quantity. The dimensions of unit quantity are

$$M^{\frac{1}{2}}L^{\frac{1}{2}}T^{-1} \times T,$$

or $$M^{\frac{1}{2}}L^{\frac{1}{2}}.$$

Unlike the unit of quantity in the electrostatic system, this unit is independent of the unit of time.

Unit *difference of potential* exists between two points in an electric conductor when one erg of work is done in transferring unit quantity from one point to the other. If Q units be transferred through a difference of potential ΔV, the work done is

$$W = Q\Delta V.$$

Unit difference of potential is, therefore, measured by $\dfrac{\text{work}}{\text{quantity}}$, and its dimensions are

$$M^{\frac{1}{2}}L^{\frac{3}{2}}T^{-2}.$$

Other dimensions in both systems are left as problems for the student. Their derivation involves the application of the general rule: Ascertain the relation which the quantities have been found to bear to each other, and hence to the fundamental quantities. Discard numerical quantities as not affecting dimensions.

For the practical purposes of electrical measurement the C.G.S. electromagnetic units are found to be of inconvenient magnitude. Multiples and sub-multiples of them have been adopted by electricians in conference as better adapted to every-day measurements. Their names and values in C.G.S. electromagnetic units are:

the ohm	$= 10^9$	C.G.S. units	of resistance.
the volt	$= 10^8$	" "	" electromotive force.
the ampere	$= 10^{-1}$	" "	" current.
the coulomb	$= 10^{-1}$	" "	" quantity.
the farad	$= 10^{-9}$	" "	" capacity.
the microfarad	$= 10^{-15}$	" "	" capacity.
the joule	$= 10^7$	" "	" work (ergs).
the watt	$= 10^7$	" "	" power.

1074. Find the conversion factor required to change potential in electromagnetic units to foot-pound units.

1075. What must be taken as the unit of force in order that currents measured in electromagnetic units may appear four times as large as now?

1076. Show that the unit of resistance is independent of the unit of mass chosen.

1077. A current measured in electromagnetic units is represented by 25. What number would represent the same current if the foot-pound-second units were used?

1078. Find the conversion factor required to change the capacity of a condenser computed when the inch is taken as the unit of length, and in electrostatic units to farads.

1079. The magnetic moment of a magnet in C.G.S. units is 1000. What would it be in foot-pound-second units?

VIBRATIONS

1080. What is meant by a vibratory motion? Does the bob of a pendulum have such motion? Does the balance wheel of a watch have such motion? State any examples of vibration which occur to you.

1081. In what ways do the motions of different particles along a clock pendulum differ? In what respects are their motions alike?

1082. What kind of motion does the end of the minute hand of a clock have? How does its motion differ from that of the hour hand? the second hand?

1083. Compare the angular velocities of the hour, minute, and second hands of a clock.

1084. An elastic ball is dropped and allowed to bound and rebound from the floor until it comes to rest. Is the motion vibratory? Draw the time and height curve. Draw the time and velocity curve approximately. Explain any peculiarities of these curves. (See falling bodies.)

1085. C and E are two balls in circular and elliptic grooves on a horizontal table. OP is a rod turning about the common center of the ellipse and circle with a uniform angular velocity, and pushing the balls around. Compare the *linear velocities* of the two balls at AA', BB', etc. Compare the average linear velocity of E with the velocity of C. The periodic time of C is 40 sec. What is that of E? Is the motion of the balls vibratory? (See Fig. 94.)

1086. If *OA'*, Fig. 94, is very small, what kind of motion will the ball moving in the ellipse approach?

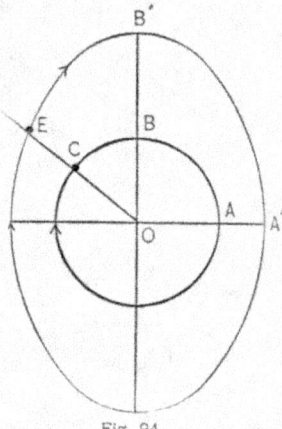

Fig. 94.

1087. How does the motion of the piston of an engine differ from that of a point in the fly-wheel?

1088. A man walks at a uniform rate in a circular track *ABCD*. Another man starts from *A* at the same time, and walks along the diameter *AC*, so that the line joining them is always perpendicular to *AC*. What kind of motion will the second man have? Where will he walk the fastest? The first goes clear around in 20 min. What is his angular velocity? What is the periodic time of the second man? Fig. 95.

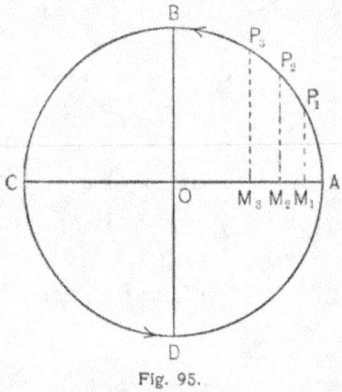

Fig. 95.

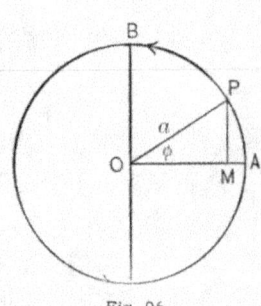

Fig. 96.

1089. If $P_1P_2 = P_2P_3$, does $M_1M_2 = M_2M_3$? The time required for the first to move from P_1 to P_2 is the same as from P_2 to P_3, and equals that for the second to go from M_1 to M_2 or M_2 to M_3. How has the motion of the second man changed in going from M_1 to M_3? Fig. 96.

If *P* moves uniformly in a circle of radius *a*, and *M* is the foot of the perpendicular dropped from *P* on a diameter *OA*, we have from trigonometry $OM = a \cos \phi$. Making all measurements from OA, and calling ω the angle turned through in 1 sec., we have $\phi = \omega t$.

Then displacement of M from center is

$$OM = x = a \cos \omega t.$$

The period is the same as that of P;

$$i.e. \quad T = \frac{2\pi}{\omega} \text{ or } \omega = \frac{2\pi}{T}.$$

$$\therefore x = a \cos \frac{2\pi}{T} t.$$

1090. When, *i.e.*, for what values of t is x a maximum? a minimum? How does the velocity of M vary?

1091. Draw a curve with time as x and distance from O as y. Draw the corresponding time-velocity curve. Draw the corresponding time-acceleration curve.

1092. Define simple harmonic motion and give several examples.

1093. Is S.H.M. a vibratory motion? Give an example of a vibratory motion which is not simple harmonic.

1094. A body has S.H.M. in a straight line. The expression for this motion is $y = 6 \sin 15\,t$. Draw to scale the representative circle. Find the periodic time; the amplitude. Find the velocity when $t = 3$ sec.

1095. The displacement of a particle is given by $y = 8 \cos 20\,t$. What is the maximum displacement? What is the maximum velocity? What is the acceleration when $y = 4$? What is the periodic time?

1096. If the angular velocity were doubled, how would the quantities in question be altered?

1097. A body of mass m vibrates with S.H.M. in a straight line. Find its average kinetic energy.

o

WAVES

In the study of wave motion, the student should bear in mind that all wave motions have certain similarities, and the examples given are mainly for the purpose of calling attention to these. It is by no means true that the actual motion of drops of water in the passage of a water wave are like the motion of air particles during the passage of a sound wave, yet the ideas of wave length, periodic time, velocity of propagation, amplitude, relation between the time required for a single particle to go through one complete series of its motions, and the distance moved by any and every wave element, etc., are common to both and enter into the consideration of every type of wave motion.

1098. A stone is dropped vertically into a pond of still water. It is observed that when ten circular crests have started outward, the outer one has a radius of 6 m. What is the wave length? If 40 sec. are required for the outer crest to acquire a radius of 5 m., what is the period?

1099. If a vertical section is made through the center of the wave system described above, draw the curve of intersection with the surface approximately. Would this curve change in form from instant to instant? Would it change in position?

1100. A system of water waves $\lambda = 1$ m., $v = 4$ m., is moving across a lake parallel to a row of fine wires 25 cm. apart. These wires, starting at a certain point, are numbered 0, 1, 2, 3, 4, 5. etc. At a given instant a crest is observed at the wire marked 0.

State (1) At which other wires crests would be found.

(2) At which other wires hollows or troughs would be found.

(3) At which other wires the water is at its natural level.

(4) At which other wires the water is at its natural level, but falling.

194

1101. When crests are observed at two wires 4 m. apart, how many crests would there be between them ? How many troughs ?

1102. Suppose that each individual particle moves in a circle, how many times would a particle go around its circle while a crest was traveling 20 m. ?

1103. A system of water waves is moving across a lake. The wave length is 5 m. The velocity of propagation is 6 m. per second. A crest is observed at a stake at a given instant. Where will that crest be in 10 sec. ? Where was it 20 sec. before ? At the instant when the crest is at the stake mentioned, what was the condition at a stake 10 m. back ? 15 m. back ? $16\frac{1}{4}$ m. back ? $17\frac{1}{2}$ m. back ? $18\frac{3}{4}$ m. back ?

1104. Two exactly similar wave systems are moving in *opposite* directions. Show by diagram how "nodes" and "loops" will be formed.

NOTE. — The student can easily trace or copy a sine curve on a card, and then cut out a pattern so as to readily draw two like curves. Then compound them by the ordinary method. Now move one $\frac{1}{4}\lambda$ to the *right* and the other the same distance to the *left*, and again compound them. Move each again, etc. It will be found that certain points will be permanently at *rest* and others vibrate with greater or less amplitude.

1105. Distinguish clearly between a progressive and a stationary wave system. Show how a stationary system may be produced.

1106. A system of progressive waves is moving in a straight line. The wave length and velocity of propagation is known and the complete history of the motion of one particle is given. What can be inferred from this ?

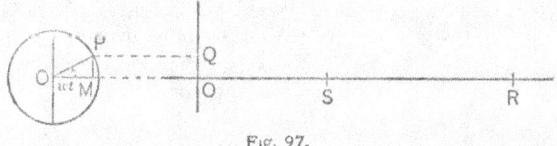

Fig. 97.

1107. A wave motion of simple harmonic type is propagated along OX (Fig. 97). The wave length is λ, the velocity of propagation is v. The circle at the left is called the **circle of reference**, which means that as P moves around the circle with uniform angular velocity the line PM, varying harmonically, is a representative of the actual motion of every disturbed particle of the medium. How far will the wave travel through the medium while P goes once around the circle?

1108. Show that $T = \dfrac{2\pi}{\omega} = \dfrac{\lambda}{v}$, where T is the common periodic time.

1109. What relation is there between the angle turned through by ρ and the distance traversed by every portion of the wave disturbance in that time?

1110. Use this relation to modify $y = a \sin \omega t$ so as to express a *progressive* wave disturbance of *simple* harmonic type.

1111. Show that $\quad y = a \sin(\omega t + \omega t')$

$$= a \sin \frac{2\pi}{T}[t + t']$$

$$= a \sin \frac{2\pi v}{\lambda}[t + t']$$

$$= a \sin \frac{2\pi}{\lambda}[vt + x]. \qquad\qquad [x = vt'.$$

1112. Show that if the displacement at S is

$$y = a \sin \frac{2\pi}{\lambda}(vt + x),$$

it is identical with that which was at the origin $\dfrac{x}{v}$ sec. before.

1113. The displacement at S is now given by

$$y = a \sin \frac{2\pi}{\lambda}(vt + x).$$

What will represent it when it reaches R, a distance l beyond? What was it represented by when at a point l units back of S?

1114. If $y = 4\sin[10\,t + 5\,x]$ is the expression for a progressive wave, what is the periodic time? the wave length? the velocity of propagation?

1115. Waves of length 2 m. pass a certain point. It is observed that four pass per second. Write the expression for their motion.

1116. From the equation $y = a \sin \dfrac{2\pi}{\lambda}(vt + x)$, we see that as t increases so that $t' - t = T = \dfrac{\lambda}{v}$, y takes all values between $+ a$ and $- a$. While if t is constant, that is, at any instant of time, all possible values for y may be found by varying x from x to $x + \lambda$. What fact does this express?

1117. How does the energy distribution of a progressive wave system differ from that of a stationary system?

1118. Two progressive wave systems, wave lengths $2:3$, are compounded. Sketch approximately the resultant in various phase relations.

1119. What do you mean by the terms *like phase, opposite phase, retardation of* $(2n + 1)\dfrac{\lambda}{2}$?

1120. Two wave systems of equal frequency are compounded. Sketch approximately the resultant wave form in the following cases :

(*a*) When the phases are alike and amplitudes equal.

(*b*) When the phases are alike and amplitudes are as $1:2$.

(*c*) When the phase difference is $45°$, and amplitudes are as $1:2$.

(*d*) When the phase difference is $90°$, and amplitudes equal.

(*e*) When the phase difference is $180°$, and amplitudes equal.

(*f*) When the phase difference is $180°$, and amplitudes $1:2$.

1121. The displacement of a point is given by $y_1 + y_2$, where

$$y_1 = A_1 \cos(\omega t + x_1),$$
$$y_2 = A_2 \cos(\omega t + x_2).$$

Find the resultant displacement, and discuss the expression obtained.

SOUND

1122. If a sounding body were in the air, and at a considerable distance from the earth, what would be the form of the wave front if the temperature were uniform? What would be the direction of motion of those air particles in the same vertical line as the source of sound? the same horizontal line? in a line at an angle of 30° with the vertical?

1123. If the velocity of sound in air is different in different directions, how would the wave form be altered?

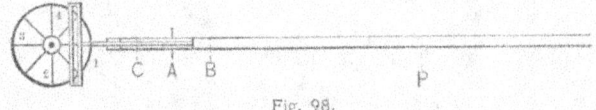

Fig. 98.

Suppose the air in an indefinitely long tube disturbed by the motion of the piston, connected as shown in Fig. 98. Let the wheel be imagined to make *one* revolution at a uniform angular velocity in the one-hundredth part of a second. When the piston reaches *B*, assume that the air at *P* is just about to be disturbed. Remembering that the disturbance will travel down the tube at a uniform velocity, draw diagrams showing the state of the air in the tube when crank pin is at 1, 2, 3, 4, indicating,

(*a*) the points of greatest, least, and average pressure,

(*b*) the places of greatest and least displacement,

(*c*) the places of greatest and least velocity of particles of air.

1124. How far would the wave travel in 1 sec. if $AP = 83$ cm.?

NOTE. — The distance AB has been neglected in comparison with AP.

1125. How far from A would the space of undisturbed air extend at the end of 1 sec., if the wheel made only *one* revolution? What is the wave length?

1126. Describe the condition of the air in tube at the end of one-twentieth of a second, if the wheel made just two revolutions and stopped.

1127. In the tube described above, consider the history of a single lamina of air at the point P when piston makes just one vibration. Draw a curve, using time in one four-hundredth of a second as x, and (a) velocity of lamina as y; (b) displacement of lamina as y; (c) density of lamina as y.

1128. How far does the wave travel when crank pin moves through an angle of 30°? 60°? 90°? 180°? 270°? What part of a wave length in each case?

1129. Consider two points in the tube a distance x apart, the velocity and displacement of the first given at a time t. How long before the second will acquire that velocity and displacement? Through what angle will crank pin move in that time?

1130. The velocity of sound at 0° C. = 33240 cm. per second. Find the velocity when temperature is 25° C.

1131. Show that if $V_t = V_0 \sqrt{1 + .003665\, t}$, velocity increases nearly 60 cm. per second for 1° rise in temperature.

1132. The report of a cannon is heard 10 sec. after the flash is seen. The temperature of the air is 20° C. How far was the observer from the gun?

1133. How much is the wave length of the air wave sent out by a 256 fork altered by a rise of temperature from 0° to 20°?

1134. A whistle giving 1000 vibrations per second is 156.20 m. distant. How many complete waves between it and the observer? Temperature 0° C.

1135. The flash of a gun is seen, and 20 sec. later the report is heard. The distance is known to be 6932 m. What was the temperature?

1136. Show that $\sqrt{\dfrac{E}{D}}$ has the same dimensions as a velocity.

1137. Apply the formula to the case of iron, taking the value of Young's modulus as $18 \cdot 10''$; density 7.67.

1138. Find the ratio of the velocity of sound in brass to that in iron.

1139. A string makes 256 complete vibrations per second. When the velocity of sound is 34600 cm. per second, what is the wave length of the sound?

1140. If the temperature of the air were increased, what quantities referred to in Example 1139 would be altered?

1141. A tuning-fork makes 1024 vibrations in a second; the wave length of the sound in air is found to be 32 cm. Find the velocity of sound.

1142. Name three ways in which musical sounds differ, and explain the cause of differences.

1143. Define pitch; timbre or character.

1144. Explain the connection between the pattern developed in the "Chladni" plates and the character of the sound produced.

1145. Explain what is meant by the term *tempered scale.* What is a musical interval?

1146. Taking 256 as C, find the frequency of the notes of the major scale. (*a*) Natural scale; (*b*) when equally tempered.

STRINGS

Formula :
$$n = \frac{1}{2 \text{ length}} \sqrt{\frac{\text{stretching force}}{\text{mass per unit length}}} \; ;$$

$$i.e. \; n = \frac{1}{2\,l} \sqrt{\frac{F}{m}}.$$

Since mass per unit length = area of cross-section × density ;

$$\therefore \; n = \frac{1}{2\,l} \sqrt{\frac{F}{\text{area of cross-section}} \cdot \frac{1}{\text{density}}}$$

$$= \frac{1}{2\,l} \sqrt{\frac{T}{\delta}}.$$

$[T = \text{force per unit area of cross-section}.$

NOTE. — The mode of vibration considered above is the fundamental. The string may vibrate in any integer multiple of this number, or in combinations of such multiples.

1147. Under certain conditions of tension and length a string makes 256 complete vibrations a second. How many would it make if its length were doubled? if its tension were doubled? if its mass were doubled without making it less flexible?

1148. It is required to raise the pitch of a certain string from C to D; *i.e.* so that it shall make 9 vibrations in the same time now required for 8. In what ways might this be done? Explain.

1149. A string making 400 vibrations per second has its length and stretching force each divided by 4, and its mass per unit length multiplied by 4. What effect on the pitch if the string is made no less flexible?

1150. A wire, 1 m. of which weighs 1 g. and is 80 cm. long, is made to vibrate in unison with fork $n = 128$. What force is used to stretch it?

1151. Why is the base string of a guitar wound with fine wire? If the wire makes each centimeter of the string four times as heavy, how will the number of vibrations be altered? What objection is there to lowering the pitch by increasing the radius of the string?

1152. Explain why it is often more desirable to shorten all the strings on a banjo by means of a clamp in order to raise the pitch rather than to increase the tension of the strings.

1153. Draw a diagram to scale, showing the relative positions of the frets on a finger-board to produce the major scale.

1154. Explain how the violin illustrates the laws of transverse vibrations of strings.

1155. What length of steel wire, mass of 1 m. $= .98$ g., stretching force weight of 9 kg. ($g = 980$), will make 256 complete vibrations per second?

$$256 = \frac{1}{2l} \sqrt{\frac{9000 \cdot 980}{.0098}},$$

$$L = \frac{1}{512} \sqrt{\frac{9 \cdot 98 \cdot 10^4}{98 \cdot 10^{-4}}}$$

$$= \frac{1}{512} \sqrt{9 \cdot 10^8} = \frac{30000}{512} \doteq 58.6 \text{ cm.}$$

1156. Two steel wires, mass of 1 m., respectively .98 and .45, are stretched side by side. The length of the larger is observed to be two-thirds that of smaller. Compare the forces stretching them; (*a*) when in unison; (*b*) when the smaller gives the octave of the larger.

1157. What proportional lengths of the two wires above must be taken such that when stretched with equal forces they will vibrate in unison?

1158. What proportional stretching forces will make the frequency of the smaller four-thirds that of the larger, their lengths being equal?

1159. Show that the expression for *n* is consistent with the laws of motion.

1160. Show that each form of equation given above is of proper dimensions.

1161. Two strings are carefully tuned so as to vibrate in unison in the fundamental. Will their overtones be harmonious?

1162. A long string is stretched between two rigid posts; a small portion is distorted as shown in diagram. When sud-

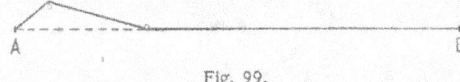

Fig. 99.

denly released it is found that triangular portion retains its shape and moves along the cord at a uniform velocity. Draw diagrams showing what happens at *B*.

1163. A uniform stretched wire is distorted as shown, *A* and *B* being rigidly fixed. The distorted portion retains its form and moves along the cord at a uniform velocity. Draw diagrams showing reflection at *D*.

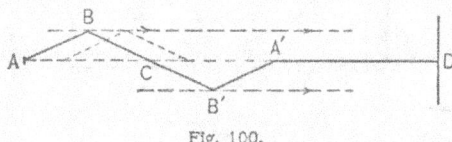

Fig. 100.

1164. Two like distortions are moving in opposite directions, and with the same velocity along a string as shown. Draw a series of diagrams showing their positions at several successive short intervals of time. Explain why the point (*P*) midway between 3 and 4 remains at rest (Fig. 101).

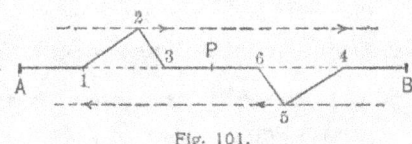

Fig. 101.

1165. Show by diagram how a string may vibrate in various modes at the same time.

STRINGS GENERAL

It is shown in books on acoustics that the equation of motion for an elastic string executing small free vibrations about a position of equilibrium is

$$m \frac{d^2y}{dt^2} = F \frac{d^2y}{dx^2},$$ (Fig. 102)

where
$$m = \text{mass per unit length,}$$
$$F = \text{stretching force,}$$
$$y = \text{displacement of a point } x \text{ distant from the origin,}$$
$$\text{at a time } t$$

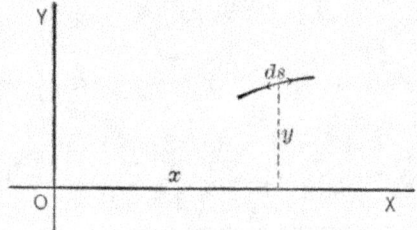

Fig. 102.

(1) Show that the equation is of consistent dimensions.

(2) Writing the equation in the form

$$\frac{d^2y}{dt^2} = \frac{F}{m} \frac{d^2y}{dx^2} = a^2 \frac{d^2y}{dx^2}, \quad \left[\frac{F}{m} = a^2,\right.$$

show by substituting that a possible relation between y, a, x, and t is

$$y = A \sin px \cos pat. \quad [A \text{ independent of } x, y, t.$$

(3) If the string is fastened at the point $x = 0$ and also at the point $x = l$ (*i.e.* at those points $y = 0$ for *all* values of t), find the least value of p.

SUGGESTION. — Sin $pl = 0$. Hence what set of values may pl have.

(4) Any part of the string between $x = 0$ and $x = l$, in other words, any point of the string free to move, will have what kind of motion?

(5) If $p = \dfrac{\pi}{l}$, what is the frequency?

(6) What other frequencies may occur? What are the tones due to these called? Is "A" the same for all of these frequencies?

(7) Does the solution given correspond to a displacement when $t = 0$, or to an initial velocity?

(8) Show that

$$y = \begin{cases} B \sin px \sin pat \\ C \cos px \cos pat \\ D \cos px \sin pat \end{cases}$$

each satisfy the original equation, and that the sum of any number of such terms is also a solution.

(9) Would the last two be consistent with a fixed point at $x = 0$?

(10) If $y = B \sin px \sin pat$ is a consistent solution, and the point $x = \dfrac{l}{3}$ were touched lightly, what would happen?

1166. Draw diagrams showing places of maximum and of minimum pressure changes in an open pipe: (*a*) when vibrating in its fundamental mode; (*b*) for the first overtone; (*c*) for the third overtone.

1167. Do the same for maximum and minimum displacements.

1168. Draw similar diagrams for a closed tube.

1169. An open pipe is vibrating in its fundamental mode; a hole in its side large enough to allow considerable air to pass in or out is suddenly opened. If the hole is at the middle of the tube, what effect will be produced?

1170. If the end of the pipe in Example 1169 is closed and the hole left open, what differences will be observed?

1171. Distinguish between "flue" and "reed" pipes, and name instruments of each class.

1172. A closed organ pipe is 60 cm. long. What is the wave length of its fundamental?

1173. What is the wave length of its first overtone?

1174. What is the wave length of the fourth overtone?

1175. When the velocity of sound in air is 34800 cm., what is the number of vibrations per second in each of the above cases?

1176. Would increase of temperature change the pitch of an organ pipe?

1177. An open tube is 100 cm. long. Find the wave length and frequency when the velocity of sound is 34000 cm. per second.

1178. What is the wave length and frequency of its first three overtones?

1179. A fork making 332 vibrations per second is fixed in front of a cylindrical tube, and the length adjusted to resonance when temperature is 0°. How much must the length be altered to resound at 25°?

1180. A closed pipe is made just long enough to reinforce a fork at its mouth, frequency of the fork 64. What must be the frequencies of the next four forks of higher pitch which it will also reinforce?

1181. What would they be if tube were open?

1182. A whistle making 4000 vibrations per second is moved slowly away from a wall. What is the first position of reinforcement? the second?

1183. How far will the whistle be from the wall when there are four nodes between it and the wall, and the sound is reinforced?

1184. How many beats per second will be heard when two forks make 250 and 255 vibrations per second respectively?

1185. How could you determine, if 6 beats per second were heard, which fork was the higher in pitch?

1186. Show by diagram how the wave giving beats is made up of two differing slightly in frequency and wave length.

1187. Explain the fluctuations in the intensity of sound from a tuning-fork when it is rotated near the ear.

1188. What are the conditions in order that two sound waves may produce silence at a point?

1189. If the scale in König's apparatus for the determination of the velocity of sound in air is 40 cm., what would be the lowest pitch which could be used as a source? For what pitch would there be found just three points where the flame was stationary?

1190. A tuning-fork making 3000 vibrations per second is slowly moved away from a wall. The velocity of sound is 34000 cm. per second. How far from the wall to the first point of resonance? to the second? to the thirteenth?

1191. Is there any difference in quality of sounds from open and closed pipes of the same fundamental pitch? If so, explain the cause.

1192. Three shortest possible tubes containing respectively air, oxygen, and hydrogen, velocities of sound, 33200, 31700, 126900, resound to a fork giving 1000 vibrations per second. What are their lengths?

1193. A locomotive whistle makes 1000 vibrations per second. When moving 50 km. per hour, what will be the alteration in pitch when approaching the observer? when receding? Temperature of air 0° C.

1194. A locomotive whistle makes 3000 vibrations per second. Find the apparent number of vibrations:

(*a*) When approaching the station at the rate of 100 km. per hour.

(*b*) When at rest and the observer is approaching the train at the same rate.

(*c*) When they are moving away from each other each at the rate of 100 km. per hour.

1195. Draw a diagram showing the effect of motion of the source relative to the air upon the wave length in air.

1196. Indicate clearly the difference between motion of the source when observer is at rest and motion of observer when source is at rest.

LIGHT

REFLECTION

1197. State the laws of reflection of light.

1198. Show how reflection is explained on the wave theory.

1199. If a mirror were perfect, could it be seen?

1200. Indicate how the form of a reflected wave front may be found when the form of the incident wave and of the reflecting surface is known.

1201. An object is placed in front of a plane mirror. Show by diagram the path of the rays by which the image is seen. What relation is there between the size of the object and the size of the image?

1202. A plane mirror is used to reflect a beam of parallel light. The mirror is turned 10°. Through what angle is the reflected beam turned? Give diagram.

1203. Show that the image formed by a plane mirror appears to be as far back of the mirror as the object is in front.

1204. Show how spherical waves reflected at a plane surface have their curvature reversed.

1205. Two mirrors are placed at an angle of 90°, with a candle between them. How many images will be seen? Locate them.

1206. If a wave after reflection is to converge to a point, what must be the wave form?

1207. Two mirrors are inclined at any angle, and a luminous point is placed between them. Show that all the images are on a circle, and determine its radius and center. Show how to find the angular position of each image.

1208. Two plane mirrors are placed parallel to each other, and 50 cm. apart. An object is placed 20 cm. from one of them. Show how the images will be spaced. Draw the path of the rays by which the fourth image on one side is seen.

1209. Explain why it is difficult to read the image of a printed page in a plane mirror.

1210. A printed sheet is laid on a table between two parallel, vertical, plane mirrors. Which of the images are easily read?

1211. A train of mirrors are placed vertical, and inclined to each other. Given the angle of incidence on the first, and the angle between the planes of each of the mirrors, find the deviation after successive reflection from each.

1212. The walls of a rectangular room are plane mirrors. A candle is placed at any point in the room, and a person standing at a given point, with his eye in the same horizontal plane as the candle, wishes to observe it by rays reflected in succession from each of the walls. Find the point at which he must look. Find the apparent distance of the image seen.

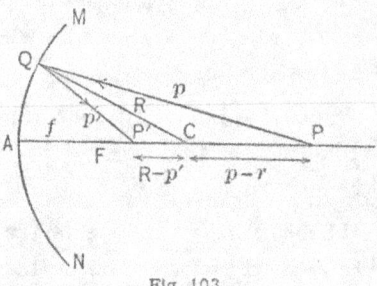

Fig. 103.

Notation used in problems relating to spherical mirrors (Fig. 103).

C . . . . center of curvature.
MN . . . aperture of mirror.
A . . . . vertex of mirror.
P . . . . luminous point.
Q . . . . point of incidence.
$CA = R$. radius of curvature.
P' . . . . intersection of reflected ray and PA.
F . . . . principal focus.

Lengths to the right from A are taken $+$.
$AP' = p' =$ image distance $= P'Q$ approximately.
$AP = p =$ object distance $= PQ$ approximately.
$AF = f =$ principal focal distance.

1213. Derive the formula showing the relation between p, p', and R.

1214. What is meant by the term *principal focus?*

1215. The radius of a concave spherical mirror is 20 cm. The sun's rays fall normally on a small portion of its surface. How far from the mirror will the image of the sun be formed?

1216. If $R = 20$ cm., find p' when $p = 40$ cm.; 35; 25; 20; 15; 12; 10; 8; 5.

For which values of p above will a real image be formed?

1217. If the object is an arrow 5 cm. high, find the size of the image in each of the cases of Example 1206. (Size refers to linear dimensions.)

1218. Construct the image as formed by a concave mirror when $p > R$, $f < p < R$, $p < f$. When is it real? when virtual? when larger than the object? when smaller?

1219. Show by diagram that if the aperture of a concave mirror is large the image formed will be distorted.

1220. With a given concave mirror where must an object be placed so that the image may be real and twice as large as the object? virtual and three times as large as the object?

1221. What must be the radius of a concave spherical mirror that an image of an object 20 ft. from a screen may be projected on the screen and be magnified three times, the object being placed between the mirror and the screen?

1222. Show how to find the position and size of the image formed by a convex mirror: (1) geometrically, (2) analytically.

1223. Derive the formula for a convex mirror, stating clearly the approximations made.

1224. A convex mirror $R = 80$ cm. is placed 30 cm. from a candle flame. Where will the image appear to be? Construct it. Find its size if the flame is 1 in. high.

1225. An object is moved from a point very near a convex mirror to a great distance away from it. How far does the image move? How would its size change?

1226. The radius of curvature of a concave mirror is 9 cm.; an object is 10 cm. in front of it. If the mirror is flattened out, *i.e.* if r increases to ∞, trace the changes in size and position of the image, neglecting the decrease of p.

1227. The radius of curvature $= 100$ cm. The object is 90 cm. from the mirror and is moving outward with a velocity of 10 cm. per second. How fast is the image moving and in which direction?

1228. A luminous point is placed at the focus of a parabolic mirror. Find the path of the reflected rays. Find the form of the wave front.

1229. Can a very small element of any wave surface be considered as spherical? If so, what would the center of the sphere mean? What surface would the center of the sphere trace as the surface element moved over the surface of the wave?

1230. State the laws of refraction. Show by diagram what you mean by the terms used in stating the law.

1231. Derive the "sine law" from consideration of the velocity of propagation of waves in the two media.

1232. If the velocity of light is altered in passing from one medium to another, does the frequency change? Does the wave length change?

1233. Does the index of refraction vary with the wave length?

1234. Show by diagram the path of a ray when passing from water to air at angles of incidence less than the critical angle; just at this angle.

1235. What is the critical angle for glass to air, index $^a\mu_g = \frac{3}{2}$?

1236. If the angle of incidence is observed to be 20° and of refraction 15°, find the index of refraction from each substance to the other.

1237. If the angle of incidence is 40° and the index is $\frac{2}{3}$, find the angle of refraction.

1238. A beam of light falls on the surface of still water at an angle of 15° with the vertical. Find its direction in the water, index $^a\mu_w = \frac{4}{3}$. Illustrate by a diagram drawn to scale.

1239. If the angle of incidence is 45°; 60°; 75°; find the direction in the water.

1240. If the angle of incidence is 45° in passing from water to air, what is the direction in air?

1241. Light is incident at an angle of 50° in water and passes into air. Find path of ray.

1242. If the direction of a ray is reversed so that it passes from water to air, what will be the index?

1243. A ray passes from water to air, angle of incidence 15°. Find direction in air.

1244. Does the critical angle depend on wave length? If so, which wave lengths would you expect to have the greater critical angle?

1245. The velocity of light in air is approximately 3.10^{10} cm. per second. What is its velocity in water, $\mu = \frac{4}{3}$? What in glass, $\mu = \frac{3}{2}$? in CS_2, $\mu = 1.63$?

1246. How much longer would it take light to reach the earth from the sun if the space were filled with water, neglecting the difference in velocity in air and vacuo? Mean distance earth to sun, 148.10^6 km.

1247. A plate of glass is immersed in water with its surface horizontal. Light is incident at an angle of 60° on the surface of the water. Find its direction in the glass, $^a\mu_w = \frac{4}{3}$, $^a\mu_g = \frac{3}{2}$.

1248. The index from air to glass is 1.5. The index from air to CS_2 is 1.6. Find the index from glass to CS_2.

1249. A beam of monochromatic light is divided; one part is sent through 1 m. of water, the other part through an air path, so that there may be no relative retardation. What is the air path required?

1250. Light is incident at an angle of 30° on a parallel plate of glass 3 cm. thick. Draw the path of the ray. How much is the beam displaced in passing through the plate, $\mu = \frac{3}{2}$?

1251. An observer estimates the depth of a pond, looking vertically downward, as 30 ft. What is the depth?

1252. If he looked from water at an object 30 ft. above the surface, how far above the surface would it appear to be?

1253. A fish is 8 ft. below the surface of the water. A man shoots at the place where the fish appears to be, holding his gun at an angle of 45° with the surface of the water. Does the bullet pass above or below the fish? (Neglect any change of direction of bullet.)

1254. Show by diagram how a straight stick held partly in water at an angle of 60° appears to a person in the air. How would it appear if the eye were under water?

1255. Under what circumstances may light be propagated in curved rather than straight lines?

1256. Explain how the sun may be seen after it has passed below the horizon.

1257. Prove that if A is the refracting angle of a prism, μ the index of refraction, δ the angle of minimum deviation,

$$\mu = \frac{\sin \frac{1}{2}(A + \delta)}{\sin \frac{1}{2}A}.$$

1258. If $A = 60°$, $\delta = 53°$, find μ.

1259. When $A = 60°$, $\mu = \frac{4}{3}$, find δ.
When $A = 30°$, $\mu = \frac{4}{3}$, find δ.

1260. Compare the minimum deviation produced by a 30° water prism and that of a similar crown-glass prism.

1261. A clear block of ice has a cavity in the form of triangular prism. The index from air to ice is 1.5. If the cavity is filled with air, show the path of a ray of light through it; if filled with a substance such that the index from ice to it were 1.6.

1262. A glass prism, index 1.5, refracting angle 60°, is placed in the path of a beam of monochromatic light. Draw a curve, using angles of incidence as abscissas and angles of deviation as ordinates.

1263. Show by diagram the path of a beam of monochromatic light passing through a glass prism placed in air; when placed in water.

1264. Show the path when white light is used.

1265. What three kinds of spectra? Explain the occurrence of dark lines in a spectrum. ('82.)

1266. Describe the experiment of the reversal of the sodium lines. What inference is drawn from this experiment? What are the three classes of spectra, and to what does each owe its origin? ('88.)

1267. Show by diagram why a slit is used as a source of light when a spectrum is required.

1268. Explain how deviation can be obtained without dispersion.

THE LENS

Refraction at a spherical surface.

Let AQ be very small compared with sphere of radius R,

P be source of light,

P_1 apparent source to an eye is second medium,

$PQ \doteq p = PA$, $\qquad \angle PQC = i$,

$P_1Q \doteq p_1 = P_1A$, $\qquad \angle P_1QC = r$.

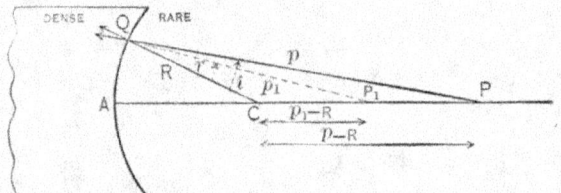

Fig. 104.

The $\triangle\, PCQ$ and P_1CQ have a common angle C.

$$\frac{\sin i}{\sin C} = \frac{p - R}{p}.$$

$$\frac{\sin C}{\sin r} = \frac{p_1}{p_1 - R}.$$

Law of sines.

$$\frac{\sin i}{\sin r} = \mu = \frac{p_1}{p_1 - R} \cdot \frac{p - R}{p};$$

i.e. $$\mu[pp_1 - pR] = pp_1 - p_1R,$$

or $$[\mu - 1]\,pp_1 = \mu pR - p_1R.$$

$$\therefore \frac{\mu - 1}{R} = \frac{\mu}{p_1} - \frac{1}{p}. \qquad (A)$$

(A) may be used to derive the formula for a lens if care is taken to observe:

(1) The index from first medium to the second is the reciprocal of the index from second to first.

(2) Distances to right are $+$, to left $-$.

216

(3) The thickness of the lens may be neglected.

(4) p_1 should be eliminated between the expressions for refraction in and out.

For example, the biconvex lens, radii R_1, R_2 (A) becomes

$$\frac{\mu - 1}{- R_1} = \frac{\mu}{p_1} - \frac{1}{p} \; \text{in,}$$

$$\frac{\frac{1}{\mu} - 1}{R_2} = \frac{\mu}{p'} - \frac{1}{p_1} \; \text{out.} \quad \left[\begin{array}{c} \text{Since } p_1 \text{ is the virtual} \\ \text{source.} \end{array} \right.$$

$$\therefore \; -(\mu - 1)\left[\frac{1}{R_1} + \frac{1}{R_2}\right] = \frac{1}{p'} - \frac{1}{p}. \quad \left[\begin{array}{c} \text{Multiply second by } \mu \text{ and} \\ \text{add the equations.} \end{array} \right.$$

If p' is negative, we have a real image or the light converges, and, changing the signs,

$$(\mu - 1)\left[\frac{1}{R_1} + \frac{1}{R_2}\right] = \frac{1}{p'} + \frac{1}{p}.$$

1269. A convex lens is placed between a source of light and a screen so as to give an image of the source on the screen. How many such positions for the lens may be found? Compare the sizes of the image and object in each case.

1270. A double convex lens, the ratio of whose radii is 6 to 1, is used as a condenser for a magic lantern. When the light is at a distance of 2 in., the emerging rays are parallel. What are the radii, the material of the lens being crown glass? ('78.)

1271. A candle is l cm. from a wall. A converging lens forms an image on the wall; when moved a distance d it also forms an image. Prove that $f = \dfrac{l^2 - d^2}{4 l}$.

1272. In a lens where $\dfrac{1}{p} - \dfrac{1}{p'} = \dfrac{1}{f}$ construct the image of an object placed between lens and F; when placed beyond F.

1273. Write a rule for the construction of images in case of spherical lenses and mirrors.

1274. The focal length of a converging lens is 3 m. Find the distance from the lens (assumed thin) to the image in each of the following positions of the object: 4 m.; 5 m.; 8 m.; 10 m.; 20 m.; 1 km.; 3 m.; 2 m.; 1 m.; 5 cm.

1275. Show by construction the position and size of the image when $f = 1$ m.; $p = 3$ m.; $p = 2$ m.; $p = .5$ m.

1276. In the derivation of the formulæ for lenses, what assumptions are made which are only approximately correct?

1277. What do you mean by a converging lens? by a diverging lens?

1278. Assuming that a biconvex lens gives a real image, construct it, and assuming that the lens is thin, prove that $\frac{1}{p} + \frac{1}{p'} = \frac{1}{f}$ by use of similar triangles.
Show also that $\dfrac{\text{size of image}}{\text{size of object}} = \dfrac{p'}{p}$

1279. By means of the formula A,
Find the formula for a biconcave lens.
Find the formula for a plano-convex lens.
Find the formula for a plano-concave lens.
Find the formula for a concavo-convex lens.

1280. Find the focal length of a biconvex lens of crown glass, $\mu = \frac{3}{2}$, $r_1 = r_2 = 30$ cm.

1281. A lens of focal length 25 in air, $^a\mu_g = \frac{3}{2}$. What will be the focal length in water, $^a\mu_w = \frac{4}{3}$.

1282. A plano-convex lens is to be made of glass, index 1.6, so as to form a real image of an object placed 2 cm. from it, and magnify it three times. What must be the radius of curvature?

1283. Find the optical center for several lenses, as biconvex of equal radii, plano-convex, etc.

1284. If q and q' are the distances of object and image from the principal focus, show that $qq' = f^2$.

1285. The radii of curvature of a biconvex lens are 30 and 32 cm. The focal length is 31 cm. What is the index of the glass?

1286. If $\mu = \frac{3}{2}$, and the radii of curvature of the biconvex lens are equal, find f.

1287. Show by diagram what you mean by chromatic aberration of a lens.

1288. Distinguish between chromatic and spherical aberration.

1289. What is meant by achromatism? How construct an achromatic lens? (Spring '79.)

1290. If values of $\frac{1}{p}$ and $\frac{1}{p'}$ are taken as co-ordinates, what kind of a curve will be found? Interpret its intercepts.

1291. If corresponding values of p and p' are measured along two rectangular lines, and p_1, p_1', p_2, p_2', etc., are joined by straight lines, show that all of these lines intersect in a point, the co-ordinates of which are $x = y = F$. (A practical fact.)

1292. If a series of observed values of p and p' are taken as abscissas and ordinates, what kind of a curve will be found?

1293. To what does the other branch of the curve correspond?

1294. A small object is placed slightly beyond the principal focus of a biconvex lens. The image formed is viewed through a biconvex lens placed nearer to the image than the principal focal distance. What is such an arrangement called? Draw a diagram showing formation of the image seen, and find the ratio of its height to that of the object.

1295. Draw diagrams showing what is meant by "short" sight or myopia. What form of lens is needed to correct myopic vision?

1296. What is meant by "long" sight, and how may it be corrected?

1297. A person is unable to see clearly objects 30 cm. from the eye. Give two possible explanations of this.

1298. Indicate by diagram how inability to decrease the radius of curvature of the crystalline lens would affect vision. What kind of glasses would be needed?

INTERFERENCE

1299. What must be the relation between the elements of two light waves in order that interference may be possible ?

1300. Explain three general methods by which interference may be obtained.

1301. Find the effective retardation of a ray of light reflected from B over one reflected from C.   Fig. 105.

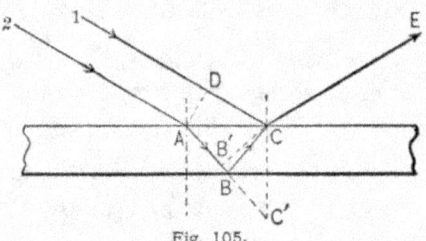

Fig. 105.

Consider parallel rays incident at A and C such that the ray refracted at A, reflected at B, and refracted at C proceeds along the same path CE as the ray reflected at C.   When 2 strikes the surface, the phase is the same as at D in 1. Draw CB' perpendicular to AB.   Then 1 travels from D to C, while 2 travels from A to B'.

Apparent retardation is $B'B + BC$.

Extend AB to C', making $BC' = BC$.

Then $BB' + CB = \delta,$

$$CC' = 2\,e.$$

$$\therefore \delta = 2\,e \cos r. \quad \text{[Retardation due to glass path.}$$

$$\therefore \delta = 2\,\mu e \cos r. \quad \text{[Equivalent retardation in air.}$$

But one reflection is with change of phase.

$\therefore$ effective retardation,

$$\delta = 2\,\mu e \cos r + \frac{\lambda}{2}.$$

It follows that if white light is reflected as shown in the figure, light of wave length λ will be a minimum when $2\,\mu e \cos r = n\lambda$.   ($n$ any integer.)

1302. What is the least thickness of crown glass, index $\frac{4}{3}$, which will give interference for sodium light when $r = 45°$?

1303. What thickness of a film, index $\frac{3}{2}$, would retard light of wave length $76 \cdot 10^{-6}$ three wave lengths?

1304. Explain the changing colored bands seen when white light is reflected from a soap-bubble film stretched vertically.

1305. What shape would the bands have if the film was attached to a ring held horizontally?

1306. White light falls on a thin wedge-shaped film of air and is reflected from each surface. It is observed that no light of wave length λ appears to come from a line parallel to the edge of the wedge and 2 mm. from the edge. Show the position of the next three lines of the same color.

1307. Explain the production of color in the soap-bubble. How can the wave length of light be measured? Derive the formula. Give diagram of apparatus used in projecting these colors on a screen. ('88.)

1308. Derive the formula for "Newton's rings."

$$\rho = \sqrt{R \sec r \cdot (2\,n+1)\dfrac{\lambda}{2}} \text{ for bright ring.}$$

$$\rho = \sqrt{R \sec r \cdot n\lambda} \qquad \text{for dark ring.}$$

1309. If red light $\lambda = 76 \cdot 10^{-6}$ is used and $R = 9$ cm., $r = 45°$, find the radii of the first four bright rings.

1310. What would be the ratio of the radii of rings of the same order for $\lambda = 76 \cdot 10^{-6}$ and $\lambda = 52 \cdot 10^{-6}$?

1311. Find the general expression for the width of the rings for a given wave length. Do they increase or decrease in width as r is increased?

DIFFRACTION

1312. Explain why the shadow of a twig cast by an arc light on a frosty pane of glass is often fringed with color.

1313. A slit in a piece of cardboard is held close to the eye and parallel to the filament of an incandescent lamp. Explain the colored fringes observed. Are the colors pure spectral colors?

1314. White light diverging from a narrow slit falls on two parallel narrow slits very close together. Show how the appearance on a screen beyond the apertures depends on the wave length considered and on the distance between the two parallel slits.

1315. Light from a small source is divided and passes by two paths of slightly different length to a screen. Explain briefly the difference in the phenomena observed when the light is white and when it is monochromatic.

1316. Parallel rays of white light fall normally on a transmission grating and the diffracted rays are brought to a focus by a lens. Show by diagram how spectra are formed and derive the formula (Fig. 106).

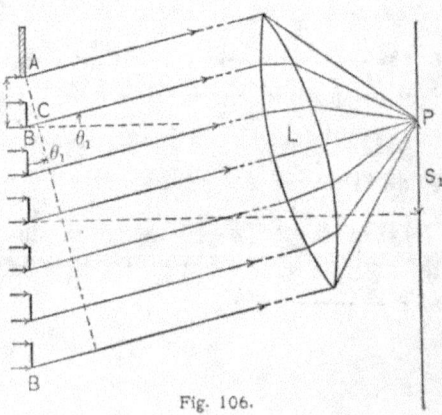

Fig. 106.

1317. Two gratings are placed one above

the other in a horizontal beam of white light from a vertical slit. If one has twice as many lines per centimeter as the other, how will the spectra differ?

1318. If $d = 10^{-3}$, $\lambda = 59 \cdot 10^{-6}$, find θ_1; θ_2; θ_3.

1319. For a certain wave length and grating, $\theta_3 = 6°$; for a different wave length, $\theta_2 = 6°$. Find the ratio of the two wave lengths and explain overlapping spectra.

1320. Show from the expression $\dfrac{n\lambda}{d} = \sin\theta_n$ how the length of the spectrum will change with d.

1321. Sunlight passing through a narrow slit falls *normally* on a transmission grating 800 lines per centimeter. The spectra are focused on a screen 10 m. from the grating. Find the position and length of the first spectrum.

1322. Light of wave length $589 \cdot 10^{-7}$ passes through the slit and falls on a grating G, Fig. 107. An eye placed just back of the grating observes a series of images of the slit, as S_1, S_2, S_3, etc. Explain how these images are formed.

If $d_1 = 5$ cm. and $l = 80$ cm., find the number of lines per centimeter in the grating.

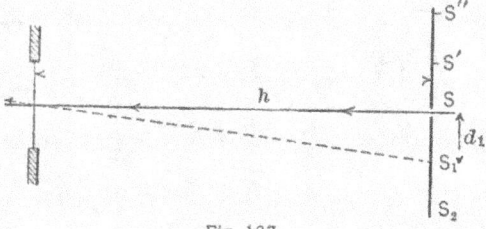

Fig. 107.

1323. How do the spectra formed by diffraction differ from those formed by refraction?

1324. What assumptions are made in the derivation of the formula for a grating which are only approximately true?

1325. Derive the formula for a reflection grating if the angle of incidence $= i$ and the grating space $= d$.

1326. Show by diagram the formation of the first spectrum by a reflection grating.

TABLES

[In these tables the admirable arrangement made use of in Bottomley's *Four-Figure Mathematical Tables* has been followed.]

	0	1	2	3	4	5	6	7	8	9	1	2	3	4	5	6	7	8	9
10	0000	0043	0086	0128	0170	0212	0253	0294	0334	0374	4	8	12	17	21	25	29	33	37
11	0414	0453	0492	0531	0569	0607	0645	0682	0719	0755	4	8	11	15	19	23	26	30	34
12	0792	0828	0864	0899	0934	0969	1004	1038	1072	1106	3	7	10	14	17	21	24	28	31
13	1139	1173	1206	1239	1271	1303	1335	1367	1399	1430	3	6	10	13	16	19	23	26	29
14	1461	1492	1523	1553	1584	1614	1644	1673	1703	1732	3	6	9	12	15	18	21	24	27
15	1761	1790	1818	1847	1875	1903	1931	1959	1987	2014	3	6	8	11	14	17	20	22	25
16	2041	2068	2095	2122	2148	2175	2201	2227	2253	2279	3	5	8	11	13	16	18	21	24
17	2304	2330	2355	2380	2405	2430	2455	2480	2504	2529	2	5	7	10	12	15	17	20	22
18	2553	2577	2601	2625	2648	2672	2695	2718	2742	2765	2	5	7	9	12	14	16	19	21
19	2788	2810	2833	2856	2878	2900	2923	2945	2967	2989	2	4	7	9	11	13	16	18	20
20	3010	3032	3054	3075	3096	3118	3139	3160	3181	3201	2	4	6	8	11	13	15	17	19
21	3222	3243	3263	3284	3304	3324	3345	3365	3385	3404	2	4	6	8	10	12	14	16	18
22	3424	3444	3464	3483	3502	3522	3541	3560	3579	3598	2	4	6	8	10	12	14	15	17
23	3617	3636	3655	3674	3692	3711	3729	3747	3766	3784	2	4	6	7	9	11	13	15	17
24	3802	3820	3838	3856	3874	3892	3909	3927	3945	3962	2	4	5	7	9	11	12	14	16
25	3979	3997	4014	4031	4048	4065	4082	4099	4116	4133	2	3	5	7	9	10	12	14	15
26	4150	4166	4183	4200	4216	4232	4249	4265	4281	4298	2	3	5	7	8	10	11	13	15
27	4314	4330	4346	4362	4378	4393	4409	4425	4440	4456	2	3	5	6	8	9	11	13	14
28	4472	4487	4502	4518	4533	4548	4564	4579	4594	4609	2	3	5	6	8	9	11	12	14
29	4624	4639	4654	4669	4683	4698	4713	4728	4742	4757	1	3	4	6	7	9	10	12	13
30	4771	4786	4800	4814	4829	4843	4857	4871	4886	4900	1	3	4	6	7	9	10	11	13
31	4914	4928	4942	4955	4969	4983	4997	5011	5024	5038	1	3	4	6	7	8	10	11	12
32	5051	5065	5079	5092	5105	5119	5132	5145	5159	5172	1	3	4	5	7	8	9	11	12
33	5185	5198	5211	5224	5237	5250	5263	5276	5289	5302	1	3	4	5	6	8	9	10	12
34	5315	5328	5340	5353	5366	5378	5391	5403	5416	5428	1	3	4	5	6	8	9	10	11
35	5441	5453	5465	5478	5490	5502	5514	5527	5539	5551	1	2	4	5	6	7	9	10	11
36	5563	5575	5587	5599	5611	5623	5635	5647	5658	5670	1	2	4	5	6	7	8	10	11
37	5682	5694	5705	5717	5729	5740	5752	5763	5775	5786	1	2	3	5	6	7	8	9	10
38	5798	5809	5821	5832	5843	5855	5866	5877	5888	5899	1	2	3	5	6	7	8	9	10
39	5911	5922	5933	5944	5955	5966	5977	5988	5999	6010	1	2	3	4	5	7	8	9	10
40	6021	6031	6042	6053	6064	6075	6085	6096	6107	6117	1	2	3	4	5	6	8	9	10
41	6128	6138	6149	6160	6170	6180	6191	6201	6212	6222	1	2	3	4	5	6	7	8	9
42	6232	6243	6253	6263	6274	6284	6294	6304	6314	6325	1	2	3	4	5	6	7	8	9
43	6335	6345	6355	6365	6375	6385	6395	6405	6415	6425	1	2	3	4	5	6	7	8	9
44	6435	6444	6454	6464	6474	6484	6493	6503	6513	6522	1	2	3	4	5	6	7	8	9
45	6532	6542	6551	6561	6571	6580	6590	6599	6609	6618	1	2	3	4	5	6	7	8	9
46	6628	6637	6646	6656	6665	6675	6684	6693	6702	6712	1	2	3	4	5	6	7	7	8
47	6721	6730	6739	6749	6758	6767	6776	6785	6794	6803	1	2	3	4	5	5	6	7	8
48	6812	6821	6830	6839	6848	6857	6866	6875	6884	6893	1	2	3	4	4	5	6	7	8
49	6902	6911	6920	6928	6937	6946	6955	6964	6972	6981	1	2	3	4	4	5	6	7	8
50	6990	6998	7007	7016	7024	7033	7042	7050	7059	7067	1	2	3	3	4	5	6	7	8
51	7076	7084	7093	7101	7110	7118	7126	7135	7143	7152	1	2	3	3	4	5	6	7	8
52	7160	7168	7177	7185	7193	7202	7210	7218	7226	7235	1	2	2	3	4	5	6	7	7
53	7243	7251	7259	7267	7275	7284	7292	7300	7308	7316	1	2	2	3	4	5	6	6	7
54	7324	7332	7340	7348	7356	7364	7372	7380	7388	7396	1	2	2	3	4	5	6	6	7

	0	1	2	3	4	5	6	7	8	9	1	2	3	4	5	6	7	8	9
55	7404	7412	7419	7427	7435	7443	7451	7459	7466	7474	1	2	2	3	4	5	5	6	7
56	7482	7490	7497	7505	7513	7520	7528	7536	7543	7551	1	2	2	3	4	5	5	6	7
57	7559	7566	7574	7582	7589	7597	7604	7612	7619	7627	1	2	2	3	4	5	5	6	7
58	7634	7642	7649	7657	7664	7672	7679	7686	7694	7701	1	1	2	3	4	4	5	6	7
59	7709	7716	7723	7731	7738	7745	7752	7760	7767	7774	1	1	2	3	4	4	5	6	7
60	7782	7789	7796	7803	7810	7818	7825	7832	7839	7846	1	1	2	3	4	4	5	6	6
61	7853	7860	7868	7875	7882	7889	7896	7903	7910	7917	1	1	2	3	4	4	5	6	6
62	7924	7931	7938	7945	7952	7959	7966	7973	7980	7987	1	1	2	3	3	4	5	6	6
63	7993	8000	8007	8014	8021	8028	8035	8041	8048	8055	1	1	2	3	3	4	5	5	6
64	8062	8069	8075	8082	8089	8096	8102	8109	8116	8122	1	1	2	3	3	4	5	5	6
65	8129	8136	8142	8149	8156	8162	8169	8176	8182	8189	1	1	2	3	3	4	5	5	6
66	8195	8202	8209	8215	8222	8228	8235	8241	8248	8254	1	1	2	3	3	4	5	5	6
67	8261	8267	8274	8280	8287	8293	8299	8306	8312	8319	1	1	2	3	3	4	5	5	6
68	8325	8331	8338	8344	8351	8357	8363	8370	8376	8382	1	1	2	3	3	4	4	5	6
69	8388	8395	8401	8407	8414	8420	8426	8432	8439	8445	1	1	2	2	3	4	4	5	6
70	8451	8457	8463	8470	8476	8482	8488	8494	8500	8506	1	1	2	2	3	4	4	5	6
71	8513	8519	8525	8531	8537	8543	8549	8555	8561	8567	1	1	2	2	3	4	4	5	5
72	8573	8579	8585	8591	8597	8603	8609	8615	8621	8627	1	1	2	2	3	4	4	5	5
73	8633	8639	8645	8651	8657	8663	8669	8675	8681	8686	1	1	2	2	3	4	4	5	5
74	8692	8698	8704	8710	8716	8722	8727	8733	8739	8745	1	1	2	2	3	4	4	5	5
75	8751	8756	8762	8768	8774	8779	8785	8791	8797	8802	1	1	2	2	3	3	4	5	5
76	8808	8814	8820	8825	8831	8837	8842	8848	8854	8859	1	1	2	2	3	3	4	4	5
77	8865	8871	8876	8882	8887	8893	8899	8904	8910	8915	1	1	2	2	3	3	4	4	5
78	8921	8927	8932	8938	8943	8949	8954	8960	8965	8971	1	1	2	2	3	3	4	4	5
79	8976	8982	8987	8993	8998	9004	9009	9015	9020	9025	1	1	2	2	3	3	4	4	5
80	9031	9036	9042	9047	9053	9058	9063	9069	9074	9079	1	1	2	2	3	3	4	4	5
81	9085	9090	9096	9101	9106	9112	9117	9122	9128	9133	1	1	2	2	3	3	4	4	5
82	9138	9143	9149	9154	9159	9165	9170	9175	9180	9186	1	1	2	2	3	3	4	4	5
83	9191	9196	9201	9206	9212	9217	9222	9227	9232	9238	1	1	2	2	3	3	4	4	5
84	9243	9248	9253	9258	9263	9269	9274	9279	9284	9289	1	1	2	2	3	3	4	4	5
85	9294	9299	9304	9309	9315	9320	9325	9330	9335	9340	1	1	2	2	3	3	4	4	5
86	9345	9350	9355	9360	9365	9370	9375	9380	9385	9390	1	1	2	2	3	3	4	4	5
87	9395	9400	9405	9410	9415	9420	9425	9430	9435	9440	0	1	1	2	2	3	3	4	4
88	9445	9450	9455	9460	9465	9469	9474	9479	9484	9489	0	1	1	2	2	3	3	4	4
89	9494	9499	9504	9509	9513	9518	9523	9528	9533	9538	0	1	1	2	2	3	3	4	4
90	9542	9547	9552	9557	9562	9566	9571	9576	9581	9586	0	1	1	2	2	3	3	4	4
91	9590	9595	9600	9605	9609	9614	9619	9624	9628	9633	0	1	1	2	2	3	3	4	4
92	9638	9643	9647	9652	9657	9661	9666	9671	9675	9680	0	1	1	2	2	3	3	4	4
93	9685	9689	9694	9699	9703	9708	9713	9717	9722	9727	0	1	1	2	2	3	3	4	4
94	9731	9736	9741	9745	9750	9754	9759	9763	9768	9773	0	1	1	2	2	3	3	4	4
95	9777	9782	9786	9791	9795	9800	9805	9809	9814	9818	0	1	1	2	2	3	3	4	4
96	9823	9827	9832	9836	9841	9845	9850	9854	9859	9863	0	1	1	2	2	3	3	4	4
97	9868	9872	9877	9881	9886	9890	9894	9899	9903	9908	0	1	1	2	2	3	3	4	4
98	9912	9917	9921	9926	9930	9934	9939	9943	9948	9952	0	1	1	2	2	3	3	4	4
99	9956	9961	9965	9969	9974	9978	9983	9987	9991	9996	0	1	1	2	2	3	3	3	4

	0'	6'	12'	18'	24'	30'	36'	42'	48'	54'	1	2	3	4	5
0°	0000	0017	0035	0052	0070	0087	0105	0122	0140	0157	3	6	9	12	15
1	0175	0192	0209	0227	0244	0262	0279	0297	0314	0332	3	6	9	12	15
2	0349	0366	0384	0401	0419	0436	0454	0471	0488	0506	3	6	9	12	15
3	0523	0541	0558	0576	0593	0610	0628	0645	0663	0680	3	6	9	12	15
4	0698	0715	0732	0750	0767	0785	0802	0819	0837	0854	3	6	9	12	15
5	0872	0889	0906	0924	0941	0958	0976	0993	1011	1028	3	6	9	12	14
6	1045	1063	1080	1097	1115	1132	1149	1167	1184	1201	3	6	9	12	14
7	1219	1236	1253	1271	1288	1305	1323	1340	1357	1374	3	6	9	12	14
8	1392	1409	1426	1444	1461	1478	1495	1513	1530	1547	3	6	9	12	14
9	1564	1582	1599	1616	1633	1650	1668	1685	1702	1719	3	6	9	12	14
10	1736	1754	1771	1788	1805	1822	1840	1857	1874	1891	3	6	9	12	14
11	1908	1925	1942	1959	1977	1994	2011	2028	2045	2062	3	6	9	11	14
12	2079	2096	2113	2130	2147	2164	2181	2198	2215	2232	3	6	9	11	14
13	2250	2267	2284	2300	2317	2334	2351	2368	2385	2402	3	6	8	11	14
14	2419	2436	2453	2470	2487	2504	2521	2538	2554	2571	3	6	8	11	14
15	2588	2605	2622	2639	2656	2672	2689	2706	2723	2740	3	6	8	11	14
16	2756	2773	2790	2807	2823	2840	2857	2874	2890	2907	3	6	8	11	14
17	2924	2940	2957	2974	2990	3007	3024	3040	3057	3074	3	6	8	11	14
18	3090	3107	3123	3140	3156	3173	3190	3206	3223	3239	3	6	8	11	14
19	3256	3272	3289	3305	3322	3338	3355	3371	3387	3404	3	5	8	11	14
20	3420	3437	3453	3469	3486	3502	3518	3535	3551	3567	3	5	8	11	14
21	3584	3600	3616	3633	3649	3665	3681	3697	3714	3730	3	5	8	11	14
22	3746	3762	3778	3795	3811	3827	3843	3859	3875	3891	3	5	8	11	14
23	3907	3923	3939	3955	3971	3987	4003	4019	4035	4051	3	5	8	11	14
24	4067	4083	4099	4115	4131	4147	4163	4179	4195	4210	3	5	8	11	13
25	4226	4242	4258	4274	4289	4305	4321	4337	4352	4368	3	5	8	11	13
26	4384	4399	4415	4431	4446	4462	4478	4493	4509	4524	3	5	8	10	13
27	4540	4555	4571	4586	4602	4617	4633	4648	4664	4679	3	5	8	10	13
28	4695	4710	4726	4741	4756	4772	4787	4802	4818	4833	3	5	8	10	13
29	4848	4863	4879	4894	4909	4924	4939	4955	4970	4985	3	5	8	10	13
30	5000	5015	5030	5045	5060	5075	5090	5105	5120	5135	3	5	8	10	13
31	5150	5165	5180	5195	5210	5225	5240	5255	5270	5284	2	5	7	10	12
32	5299	5314	5329	5344	5358	5373	5388	5402	5417	5432	2	5	7	10	12
33	5446	5461	5476	5490	5505	5519	5534	5548	5563	5577	2	5	7	10	12
34	5592	5606	5621	5635	5650	5664	5678	5693	5707	5721	2	5	7	10	12
35	5736	5750	5764	5779	5793	5807	5821	5835	5850	5864	2	5	7	10	12
36	5878	5892	5906	5920	5934	5948	5962	5976	5990	6004	2	5	7	9	12
37	6018	6032	6046	6060	6074	6088	6101	6115	6129	6143	2	5	7	9	12
38	6157	6170	6184	6198	6211	6225	6239	6252	6266	6280	2	5	7	9	11
39	6293	6307	6320	6334	6347	6361	6374	6388	6401	6414	2	4	7	9	11
40	6428	6441	6455	6468	6481	6494	6508	6521	6534	6547	2	4	7	9	11
41	6561	6574	6587	6600	6613	6626	6639	6652	6665	6678	2	4	7	9	11
42	6691	6704	6717	6730	6743	6756	6769	6782	6794	6807	2	4	6	9	11
43	6820	6833	6845	6858	6871	6884	6896	6909	6921	6934	2	4	6	8	11
44	6947	6959	6972	6984	6997	7009	7022	7034	7046	7059	2	4	6	8	10

	0′	6′	12′	18′	24′	30′	36′	42′	48′	54′	1	2	3	4	5
45°	7071	7083	7096	7108	7120	7133	7145	7157	7169	7181	2	4	6	8	10
46	7193	7206	7218	7230	7242	7254	7266	7278	7290	7302	2	4	6	8	10
47	7314	7325	7337	7349	7361	7373	7385	7396	7408	7420	2	4	6	8	10
48	7431	7443	7455	7466	7478	7490	7501	7513	7524	7536	2	4	6	8	10
49	7547	7558	7570	7581	7593	7604	7615	7627	7638	7649	2	4	6	8	9
50	7660	7672	7683	7694	7705	7716	7727	7738	7749	7760	2	4	6	7	9
51	7771	7782	7793	7804	7815	7826	7837	7848	7859	7869	2	4	5	7	9
52	7880	7891	7902	7912	7923	7934	7944	7955	7965	7976	2	4	5	7	9
53	7986	7997	8007	8018	8028	8039	8049	8059	8070	8080	2	3	5	7	9
54	8090	8100	8111	8121	8131	8141	8151	8161	8171	8181	2	3	5	7	8
55	8192	8202	8211	8221	8231	8241	8251	8261	8271	8281	2	3	5	7	8
56	8290	8300	8310	8320	8329	8339	8348	8358	8368	8377	2	3	5	6	8
57	8387	8396	8406	8415	8425	8434	8443	8453	8462	8471	2	3	5	6	8
58	8480	8490	8499	8508	8517	8526	8536	8545	8554	8563	2	3	5	6	8
59	8572	8581	8590	8599	8607	8616	8625	8634	8643	8652	1	3	4	6	7
60	8660	8669	8678	8686	8695	8704	8712	8721	8729	8738	1	3	4	6	7
61	8746	8755	8763	8771	8780	8788	8796	8805	8813	8821	1	3	4	6	7
62	8829	8838	8846	8854	8862	8870	8878	8886	8894	8902	1	3	4	5	7
63	8910	8918	8926	8934	8942	8949	8957	8965	8973	8980	1	3	4	5	6
64	8988	8996	9003	9011	9018	9026	9033	9041	9048	9056	1	3	4	5	6
65	9063	9070	9078	9085	9092	9100	9107	9114	9121	9128	1	2	4	5	6
66	9135	9143	9150	9157	9164	9171	9178	9184	9191	9198	1	2	3	5	6
67	9205	9212	9219	9225	9232	9239	9245	9252	9259	9265	1	2	3	4	6
68	9272	9278	9285	9291	9298	9304	9311	9317	9323	9330	1	2	3	4	5
69	9336	9342	9348	9354	9361	9367	9373	9379	9385	9391	1	2	3	4	5
70	9397	9403	9409	9415	9421	9426	9432	9438	9444	9449	1	2	3	4	5
71	9455	9461	9466	9472	9478	9483	9489	9494	9500	9505	1	2	3	4	5
72	9511	9516	9521	9527	9532	9537	9542	9548	9553	9558	1	2	3	4	4
73	9563	9568	9573	9578	9583	9588	9593	9598	9603	9608	1	2	2	3	4
74	9613	9617	9622	9627	9632	9636	9641	9646	9650	9655	1	2	2	3	4
75	9659	9664	9668	9673	9677	9681	9686	9690	9694	9699	1	1	2	3	4
76	9703	9707	9711	9715	9720	9724	9728	9732	9736	9740	1	1	2	3	3
77	9744	9748	9751	9755	9759	9763	9767	9770	9774	9778	1	1	2	3	3
78	9781	9785	9789	9792	9796	9799	9803	9806	9810	9813	1	1	2	2	3
79	9816	9820	9823	9826	9829	9833	9836	9839	9842	9845	1	1	2	2	3
80	9848	9851	9854	9857	9860	9863	9866	9869	9871	9874	0	1	1	2	2
81	9877	9880	9882	9885	9888	9890	9893	9895	9898	9900	0	1	1	2	2
82	9903	9905	9907	9910	9912	9914	9917	9919	9921	9923	0	1	1	2	2
83	9925	9928	9930	9932	9934	9936	9938	9940	9942	9943	0	1	1	1	2
84	9945	9947	9949	9951	9952	9954	9956	9957	9959	9960	0	1	1	1	1
85	9962	9963	9965	9966	9968	9969	9971	9972	9973	9974	0	0	1	1	1
86	9976	9977	9978	9979	9980	9981	9982	9983	9984	9985	0	0	1	1	1
87	9986	9987	9988	9989	9990	9990	9991	9992	9993	9993	0	0	0	1	1
88	9994	9995	9995	9996	9996	9997	9997	9997	9998	9998	0	0	0	0	0
89	9998	9999	9999	9999	9999	1·000 nearly.	1·000 nearly.	1·000 nearly.	1·000 nearly.	1·000 nearly.	0	0	0	0	0

NATURAL COSINES

	0'	6'	12'	18'	24'	30'	36'	42'	48'	54'	1	2	3	4	5
0°	1·000	1·000 nearly.	1·000 nearly.	1·000 nearly.	1·000 nearly.	9999	9999	9999	9999	9999	0	0	0	0	0
1	9998	9998	9998	9997	9997	9997	9996	9996	9995	9995	0	0	0	0	0
2	9994	9993	9993	9992	9991	9990	9990	9989	9988	9987	0	0	0	1	1
3	9986	9985	9984	9983	9982	9981	9980	9979	9978	9977	0	0	1	1	1
4	9976	9974	9973	9972	9971	9969	9968	9966	9965	9963	0	0	1	1	1
5	9962	9960	9959	9957	9956	9954	9952	9951	9949	9947	0	1	1	1	2
6	9945	9943	9942	9940	9938	9936	9934	9932	9930	9928	0	1	1	1	2
7	9925	9923	9921	9919	9917	9914	9912	9910	9907	9905	0	1	1	2	2
8	9903	9900	9898	9895	9893	9890	9888	9885	9882	9880	0	1	1	2	2
9	9877	9874	9871	9869	9866	9863	9860	9857	9854	9851	0	1	1	2	2
10	9848	9845	9842	9839	9836	9833	9829	9826	9823	9820	1	1	2	2	3
11	9816	9813	9810	9806	9803	9799	9796	9792	9789	9785	1	1	2	2	3
12	9781	9778	9774	9770	9767	9763	9759	9755	9751	9748	1	1	2	3	3
13	9744	9740	9736	9732	9728	9724	9720	9715	9711	9707	1	1	2	3	3
14	9703	9699	9694	9690	9686	9681	9677	9673	9668	9664	1	1	2	3	4
15	9659	9655	9650	9646	9641	9636	9632	9627	9622	9617	1	2	2	3	4
16	9613	9608	9603	9598	9593	9588	9583	9578	9573	9568	1	2	2	3	4
17	9563	9558	9553	9548	9542	9537	9532	9527	9521	9516	1	2	3	4	4
18	9511	9505	9500	9494	9489	9483	9478	9472	9466	9461	1	2	3	4	5
19	9455	9449	9444	9438	9432	9426	9421	9415	9409	9403	1	2	3	4	5
20	9397	9391	9385	9379	9373	9367	9361	9354	9348	9342	1	2	3	4	5
21	9336	9330	9323	9317	9311	9304	9298	9291	9285	9278	1	2	3	4	5
22	9272	9265	9259	9252	9245	9239	9232	9225	9219	9212	1	2	3	4	6
23	9205	9198	9191	9184	9178	9171	9164	9157	9150	9143	1	2	3	5	6
24	9135	9128	9121	9114	9107	9100	9092	9085	9078	9070	1	2	4	5	6
25	9063	9056	9048	9041	9033	9026	9018	9011	9003	8996	1	3	4	5	6
26	8988	8980	8973	8965	8957	8949	8942	8934	8926	8918	1	3	4	5	6
27	8910	8902	8894	8886	8878	8870	8862	8854	8846	8838	1	3	4	5	7
28	8829	8821	8813	8805	8796	8788	8780	8771	8763	8755	1	3	4	6	7
29	8746	8738	8729	8721	8712	8704	8695	8686	8678	8669	1	3	4	6	7
30	8660	8652	8643	8634	8625	8616	8607	8599	8590	8581	1	3	4	6	7
31	8572	8563	8554	8545	8536	8526	8517	8508	8499	8490	2	3	5	6	8
32	8480	8471	8462	8453	8443	8434	8425	8415	8406	8396	2	3	5	6	8
33	8387	8377	8368	8358	8348	8339	8329	8320	8310	8300	2	3	5	6	8
34	8290	8281	8271	8261	8251	8241	8231	8221	8211	8202	2	3	5	7	8
35	8192	8181	8171	8161	8151	8141	8131	8121	8111	8100	2	3	5	7	8
36	8090	8080	8070	8059	8049	8039	8028	8018	8007	7997	2	3	5	7	9
37	7986	7976	7965	7955	7944	7934	7923	7912	7902	7891	2	4	5	7	9
38	7880	7869	7859	7848	7837	7826	7815	7804	7793	7782	2	4	5	7	9
39	7771	7760	7749	7738	7727	7716	7705	7694	7683	7672	2	4	6	7	9
40	7660	7649	7638	7627	7615	7604	7593	7581	7570	7559	2	4	6	8	9
41	7547	7536	7524	7513	7501	7490	7478	7466	7455	7443	2	4	6	8	10
42	7431	7420	7408	7396	7385	7373	7361	7349	7337	7325	2	4	6	8	10
43	7314	7302	7290	7278	7266	7254	7242	7230	7218	7206	2	4	6	8	10
44	7193	7181	7169	7157	7145	7133	7120	7108	7096	7083	2	4	6	8	10

N.B.—Numbers in difference-columns to be subtracted, not added.

	0'	6'	12'	18'	24'	30'	36'	42'	48'	54'	1	2	3	4	5
45°	7071	7059	7046	7034	7022	7009	6997	6984	6972	6959	2	4	6	8	10
46	6947	6934	6921	6909	6896	6884	6871	6858	6845	6833	2	4	6	8	11
47	6820	6807	6794	6782	6769	6756	6743	6730	6717	6704	2	4	6	9	11
48	6691	6678	6665	6652	6639	6626	6613	6600	6587	6574	2	4	7	9	11
49	6561	6547	6534	6521	6508	6494	6481	6468	6455	6441	2	4	7	9	11
50	6428	6414	6401	6388	0374	6361	6347	6334	6320	6307	2	4	7	9	11
51	6293	6280	6266	6252	6239	6225	6211	6198	6184	6170	2	5	7	9	11
52	6157	6143	6129	6115	6101	6088	6074	6060	6046	6032	2	5	7	9	12
53	6018	6004	5990	5976	5962	5948	5934	5920	5906	5892	2	5	7	9	12
54	5878	5864	5850	5835	5821	5807	5793	5779	5764	5750	2	5	7	9	12
55	5736	5721	5707	5693	5678	5664	5650	5635	5621	5606	2	5	7	10	12
56	5592	5577	5563	5548	5534	5519	5505	5490	5476	5461	2	5	7	10	12
57	5446	5432	5417	5402	5388	5373	5358	5344	5329	5314	2	5	7	10	12
58	5299	5284	5270	5255	5240	5225	5210	5195	5180	5165	2	5	7	10	12
59	5150	5135	5120	5105	5090	5075	5060	5045	5030	5015	3	5	8	10	13
60	5000	4985	4970	4955	4939	4924	4909	4894	4879	4863	3	5	8	10	13
61	4848	4833	4818	4802	4787	4772	4756	4741	4726	4710	3	5	8	10	13
62	4695	4679	4664	4648	4633	4617	4602	4586	4571	4555	3	5	8	10	13
63	4540	4524	4509	4493	4478	4462	4446	4431	4415	4399	3	5	8	10	13
64	4384	4368	4352	4337	4321	4305	4289	4274	4258	4242	3	5	8	11	13
65	4226	4210	4195	4179	4163	4147	4131	4115	4099	4083	3	5	8	11	13
66	4067	4051	4035	4019	4003	3987	3971	3955	3939	3923	3	5	8	11	14
67	3907	3891	3875	3859	3843	3827	3811	3795	3778	3762	3	5	8	11	14
68	3746	3730	3714	3697	3681	3665	3649	3633	3616	3600	3	5	8	11	14
69	3584	3567	3551	3535	3518	3502	3486	3469	3453	3437	3	5	8	11	14
70	3420	3404	3387	3371	3355	3338	3322	3305	3289	3272	3	5	8	11	14
71	3256	3239	3223	3206	3190	3173	3156	3140	3123	3107	3	6	8	11	14
72	3090	3074	3057	3040	3024	3007	2990	2974	2957	2940	3	6	8	11	14
73	2924	2907	2890	2874	2857	2840	2823	2807	2790	2773	3	6	8	11	14
74	2756	2740	2723	2706	2689	2672	2656	2639	2622	2605	3	6	8	11	14
75	2588	2571	2554	2538	2521	2504	2487	2470	2453	2436	3	6	8	11	14
76	2419	2402	2385	2368	2351	2334	2317	2300	2284	2267	3	6	8	11	14
77	2250	2233	2215	2198	2181	2164	2147	2130	2113	2096	3	6	9	11	14
78	2079	2062	2045	2028	2011	1994	1977	1959	1942	1925	3	6	9	11	14
79	1908	1891	1874	1857	1840	1822	1805	1788	1771	1754	3	6	9	12	14
80	1736	1719	1702	1685	1668	1650	1633	1616	1599	1582	3	6	9	12	14
81	1564	1547	1530	1513	1495	1478	1461	1444	1426	1409	3	6	9	12	14
82	1392	1374	1357	1340	1323	1305	1288	1271	1253	1236	3	6	9	12	14
83	1219	1201	1184	1167	1149	1132	1115	1097	1080	1063	3	6	9	12	14
84	1045	1028	1011	0993	0976	0958	0941	0924	0906	0889	3	6	9	12	14
85	0872	0854	0837	0819	0802	0785	0767	0750	0732	0715	3	6	9	12	15
86	0698	0680	0663	0645	0628	0610	0593	0576	0558	0541	3	6	9	12	15
87	0523	0506	0488	0471	0454	0436	0419	0401	0384	0366	3	6	9	12	15
88	0349	0332	0314	0297	0279	0262	0244	0227	0209	0192	3	6	9	12	15
89	0175	0157	0140	0122	0105	0087	0070	0052	0035	0017	3	6	9	12	15

N.B. — Numbers in difference-columns to be subtracted, not added.

	0′	6′	12′	18′	24′	30′	36′	42′	48′	54′	1	2	3	4	5
0°	·0000	0017	0035	0052	0070	0087	0105	0122	0140	0157	3	6	9	12	14
1	·0175	0192	0209	0227	0244	0262	0279	0297	0314	0332	3	6	9	12	15
2	·0349	0367	0384	0402	0419	0437	0454	0472	0489	0507	3	6	9	12	15
3	·0524	0542	0559	0577	0594	0612	0629	0647	0664	0682	3	6	9	12	15
4	·0699	0717	0734	0752	0769	0787	0805	0822	0840	0857	3	6	9	12	15
5	·0875	0892	0910	0928	0945	0963	0981	0998	1016	1033	3	6	9	12	15
6	·1051	1069	1086	1104	1122	1139	1157	1175	1192	1210	3	6	9	12	15
7	·1228	1246	1263	1281	1299	1317	1334	1352	1370	1388	3	6	9	12	15
8	·1405	1423	1441	1459	1477	1495	1512	1530	1548	1566	3	6	9	12	15
9	·1584	1602	1620	1638	1655	1673	1691	1709	1727	1745	3	6	9	12	15
10	·1763	1781	1799	1817	1835	1853	1871	1890	1908	1926	3	6	9	12	15
11	·1944	1962	1980	1998	2016	2035	2053	2071	2089	2107	3	6	9	12	15
12	·2126	2144	2162	2180	2199	2217	2235	2254	2272	2290	3	6	9	12	15
13	·2309	2327	2345	2364	2382	2401	2419	2438	2456	2475	3	6	9	12	15
14	·2493	2512	2530	2549	2568	2586	2605	2623	2642	2661	3	6	9	12	16
15	·2679	2698	2717	2736	2754	2773	2792	2811	2830	2849	3	6	9	13	16
16	·2867	2886	2905	2924	2943	2962	2981	3000	3019	3038	3	6	9	13	16
17	·3057	3076	3096	3115	3134	3153	3172	3191	3211	3230	3	6	10	13	16
18	·3249	3269	3288	3307	3327	3346	3365	3385	3404	3424	3	6	10	13	16
19	·3443	3463	3482	3502	3522	3541	3561	3581	3600	3620	3	6	10	13	17
20	·3640	3659	3679	3699	3719	3739	3759	3779	3799	3819	3	7	10	13	17
21	·3839	3859	3879	3899	3919	3939	3959	3979	4000	4020	3	7	10	13	17
22	·4040	4061	4081	4101	4122	4142	4163	4183	4204	4224	3	7	10	14	17
23	·4245	4265	4286	4307	4327	4348	4369	4390	4411	4431	3	7	10	14	17
24	·4452	4473	4494	4515	4536	4557	4578	4599	4621	4642	4	7	10	14	18
25	·4663	4684	4706	4727	4748	4770	4791	4813	4834	4856	4	7	11	14	18
26	·4877	4899	4921	4942	4964	4986	5008	5029	5051	5073	4	7	11	15	18
27	·5095	5117	5139	5161	5184	5206	5228	5250	5272	5295	4	7	11	15	18
28	·5317	5340	5362	5384	5407	5430	5452	5475	5498	5520	4	8	11	15	19
29	·5543	5566	5589	5612	5635	5658	5681	5704	5727	5750	4	8	12	15	19
30	·5774	5797	5820	5844	5867	5890	5914	5938	5961	5985	4	8	12	16	20
31	·6009	6032	6056	6080	6104	6128	6152	6176	6200	6224	4	8	12	16	20
32	·6249	6273	6297	6322	6346	6371	6395	6420	6445	6469	4	8	12	16	20
33	·6494	6519	6544	6569	6594	6619	6644	6669	6694	6720	4	8	13	17	21
34	·6745	6771	6796	6822	6847	6873	6899	6924	6950	6976	4	9	13	17	21
35	·7002	7028	7054	7080	7107	7133	7159	7186	7212	7239	4	9	13	18	22
36	·7265	7292	7319	7346	7373	7400	7427	7454	7481	7508	5	9	14	18	23
37	·7536	7563	7590	7618	7646	7673	7701	7729	7757	7785	5	9	14	18	23
38	·7813	7841	7869	7898	7926	7954	7983	8012	8040	8069	5	10	14	19	24
39	·8098	8127	8156	8185	8214	8243	8273	8302	8332	8361	5	10	15	20	24
40	·8391	8421	8451	8481	8511	8541	8571	8601	8632	8662	5	10	15	20	25
41	·8693	8724	8754	8785	8816	8847	8878	8910	8941	8972	5	10	16	21	26
42	·9004	9036	9067	9099	9131	9163	9195	9228	9260	9293	5	11	16	21	27
43	·9325	9358	9391	9424	9457	9490	9523	9556	9590	9623	6	11	17	22	28
44	·9657	9691	9725	9759	9793	9827	9861	9896	9930	9965	6	11	17	23	29

	O'	6'	12'	18'	24'	30'	36'	42'	48'	54'	1	2	3	4	5
45°	1·0000	0035	0070	0105	0141	0176	0212	0247	0283	0319	6	12	18	24	30
46	1·0355	0392	0428	0464	0501	0538	0575	0612	0649	0686	6	12	18	25	31
47	1·0724	0761	0799	0837	0875	0913	0951	0990	1028	1067	6	13	19	25	32
48	1·1106	1145	1184	1224	1263	1303	1343	1383	1423	1463	7	13	20	26	33
49	1·1504	1544	1585	1626	1667	1708	1750	1792	1833	1875	7	14	21	28	34
50	1·1918	1960	2002	2045	2088	2131	2174	2218	2261	2305	7	14	22	29	36
51	1·2349	2393	2437	2482	2527	2572	2617	2662	2708	2753	8	15	23	30	38
52	1·2799	2846	2892	2938	2985	3032	3079	3127	3175	3222	8	16	23	31	39
53	1·3270	3319	3367	3416	3465	3514	3564	3613	3663	3713	8	16	25	33	41
54	1·3764	3814	3865	3916	3968	4019	4071	4124	4176	4229	9	17	26	34	43
55	1·4281	4335	4388	4442	4496	4550	4605	4659	4715	4770	9	18	27	36	45
56	1·4826	4882	4938	4994	5051	5108	5166	5224	5282	5340	10	19	29	38	48
57	1·5399	5458	5517	5577	5637	5697	5757	5818	5880	5941	10	20	30	40	50
58	1·6003	6066	6128	6191	6255	6319	6383	6447	6512	6577	11	21	32	43	53
59	1·6643	6709	6775	6842	6909	6977	7045	7113	7182	7251	11	23	34	45	56
60	1·7321	7391	7461	7532	7603	7675	7747	7820	7893	7966	12	24	36	48	60
61	1·8040	8115	8190	8265	8341	8418	8495	8572	8650	8728	13	26	38	51	64
62	1·8807	8887	8967	9047	9128	9210	9292	9375	9458	9542	14	27	41	55	68
63	1·9626	9711	9797	9883	9970	0057	0145	0233	0323	0413	15	29	44	58	73
64	2·0503	0594	0686	0778	0872	0965	1060	1155	1251	1348	16	31	47	63	78
65	2·1445	1543	1642	1742	1842	1943	2045	2148	2251	2355	17	34	51	68	85
66	2·2460	2566	2673	2781	2889	2998	3109	3220	3332	3445	18	37	55	74	92
67	2·3559	3673	3789	3906	4023	4142	4262	4383	4504	4627	20	40	60	79	99
68	2·4751	4876	5002	5129	5257	5386	5517	5649	5782	5916	22	43	65	87	108
69	2·6051	6187	6325	6464	6605	6746	6889	7034	7179	7326	24	47	71	95	118
70	2·7475	7625	7776	7929	8083	8239	8397	8556	8716	8878	26	52	78	104	130
71	2·9042	9208	9375	9544	9714	9887	0061	0237	0415	0595	29	58	87	115	144
72	3·0777	0961	1146	1334	1524	1716	1910	2106	2305	2506	32	64	96	129	161
73	3·2709	2914	3122	3332	3544	3759	3977	4197	4420	4646	36	72	108	144	180
74	3·4874	5105	5339	5576	5816	6059	6305	6554	6806	7062	41	82	122	162	203
75	3·7321	7583	7848	8118	8391	8667	8947	9232	9520	9812	46	94	139	186	232
76	4·0108	0408	0713	1022	1335	1653	1976	2303	2635	2972	53	107	160	214	267
77	4·3315	3662	4015	4374	4737	5107	5483	5864	6252	6646	62	124	186	248	310
78	4·7046	7453	7867	8288	8716	9152	9594	0045	0504	0970	73	146	219	292	365
79	5·1446	1929	2422	2924	3435	3955	4486	5026	5578	6140	87	175	262	350	437
80	5·6713	7297	7894	8502	9124	9758	0405	1066	1742	2432					
81	6·3138	3859	4596	5350	6122	6912	7920	8548	9395	0264					
82	7·1154	2066	3002	3962	4947	5958	6996	8062	9158	0285					
83	8·1443	2636	3863	5126	6427	7769	9152	0579	2052	3572					
84	9·5144	9·677	9·845	10·02	10·20	10·39	10·58	10·78	10·99	11·20					
85	11·43	11·66	11·91	12·16	12·43	12·71	13·00	13·30	13·62	13·95					
86	14·30	14·67	15·06	15·46	15·89	16·35	16·83	17·34	17·89	18·46					
87	19·08	19·74	20·45	21·20	22·02	22·90	23·86	24·90	26·03	27·27					
88	28·64	30·14	31·82	33·69	35·80	38·19	40·92	44·07	47·74	52·08					
89	57·29	63·66	71·62	81·85	95·49	114·6	143·2	191·0	286·5	573·0					

Difference-columns cease to be useful, owing to the rapidity with which the value of the tangent changes.

	0′	6′	12′	18′	24′	30′	36′	42′	48′	54′					
0°	Inf.	573·0	286·5	191·0	143·2	114·6	95·49	81·85	71·62	63·66					
1	57·29	52·08	47·74	44·07	40·92	38·19	35·80	33·69	31·82	30·14					
2	28·64	27·27	26·03	24·90	23·86	22·90	22·02	21·20	20·45	19·74					
3	19·08	18·46	17·89	17·34	16·83	16·35	15·89	15·46	15·06	14·67					
4	14·30	13·95	13·62	13·30	13·00	12·71	12·43	12·16	11·91	11·66		Difference-columns			
5	11·43	11·20	10·99	10·78	10·58	10·39	10·20	10·02	9·845	9·677		not useful here, owing			
6	9·5144	3572	2052	0579	9152	7769	6427	5126	3863	2636		to the rapidity with			
7	8·1443	0285	9158	8062	6996	5958	4947	3962	3002	2066		which the value of the			
8	7·1154	0264	9395	8548	7920	6912	6122	5350	4596	3859		cotangent changes.			
9	6·3138	2432	1742	1066	0405	9758	9124	8502	7894	7297					
10	5·6713	6140	5578	5026	4486	3955	3435	2924	2422	1929	1	2	3	4	5
11	5·1446	0970	0504	0045	9594	9152	8716	8288	7867	7453	74	148	222	296	370
12	4·7046	6646	6252	5864	5483	5107	4737	4374	4015	3662	63	125	188	252	314
13	4·3315	2972	2635	2303	1976	1653	1335	1022	0713	0408	53	107	160	214	267
14	4·0108	9812	9520	9232	8947	8667	8391	8118	7848	7583	46	93	139	186	232
15	3·7321	7062	6806	6554	6305	6059	5816	5576	5339	5105	41	82	122	163	204
16	3·4874	4646	4420	4197	3977	3759	3544	3332	3122	2914	36	72	108	144	180
17	3·2709	2506	2305	2106	1910	1716	1524	1334	1146	0961	32	64	96	129	161
18	3·0777	0595	0415	0237	0061	9887	9714	9544	9375	9208	29	58	87	115	144
19	2·9042	8878	8716	8556	8397	8239	8083	7929	7776	7625	26	52	78	104	130
20	2·7475	7326	7179	7034	6889	6746	6605	6464	6325	6187	24	47	71	95	118
21	2·6051	5916	5782	5649	5517	5386	5257	5129	5002	4876	22	43	65	87	108
22	2·4751	4627	4504	4383	4262	4142	4023	3906	3789	3673	20	40	60	79	99
23	2·3559	3445	3332	3220	3109	2998	2889	2781	2673	2566	18	37	55	74	92
24	2·2460	2355	2251	2148	2045	1943	1842	1742	1642	1543	17	34	51	68	85
25	2·1445	1348	1251	1155	1060	0965	0872	0778	0686	0594	16	31	47	63	78
26	2·0503	0413	0323	0233	0145	0057	9970	9883	9797	9711	15	29	44	58	73
27	1·9626	9542	9458	9375	9292	9210	9128	9047	8967	8887	14	27	41	55	68
28	1·8807	8728	8650	8572	8495	8418	8341	8265	8190	8115	13	26	38	51	64
29	1·8040	7966	7893	7820	7747	7675	7603	7532	7461	7391	12	24	36	48	60
30	1·7321	7251	7182	7113	7045	6977	6909	6842	6775	6709	11	23	34	45	56
31	1·6643	6577	6512	6447	6383	6319	6255	6191	6128	6066	11	21	32	43	53
32	1·6003	5941	5880	5818	5757	5697	5637	5577	5517	5458	10	20	30	40	50
33	1·5399	5340	5282	5224	5166	5108	5051	4994	4938	4882	10	19	29	38	48
34	1·4826	4770	4715	4659	4605	4550	4496	4442	4388	4335	9	18	27	36	45
35	1·4281	4229	4176	4124	4071	4019	3968	3916	3865	3814	9	17	26	34	43
36	1·3764	3713	3663	3613	3564	3514	3465	3416	3367	3319	8	16	25	33	41
37	1·3270	3222	3175	3127	3079	3032	2985	2938	2892	2846	8	16	23	31	39
38	1·2799	2753	2708	2662	2617	2572	2527	2482	2437	2393	8	15	23	30	38
39	1·2349	2305	2261	2218	2174	2131	2088	2045	2002	1960	7	14	22	29	36
40	1·1918	1875	1833	1792	1750	1708	1667	1626	1585	1544	7	14	21	28	34
41	1·1504	1463	1423	1383	1343	1303	1263	1224	1184	1145	7	13	20	26	33
42	1·1106	1067	1028	0990	0951	0913	0875	0837	0799	0761	6	13	19	25	32
43	1·0724	0686	0649	0612	0575	0538	0501	0464	0428	0392	6	12	18	25	31
44	1·0355	0319	0283	0247	0212	0176	0141	0105	0070	0035	6	12	18	24	30

N.B. — Numbers in difference-columns to be subtracted, not added.

	0'	6'	12	18'	24'	30'	36'	42'	48'	54'	1 2 3	4 5
45°	1·0	0·9965	0·9930	0·9896	0·9861	0·9827	0·9793	0·9759	0·9725	0·9691	6 11 17	23 29
46	·9657	9623	9590	9556	9523	9490	9457	9424	9391	9358	6 11 17	22 28
47	·9325	9293	9260	9228	9195	9163	9131	9099	9067	9036	5 11 16	21 27
48	·9004	8972	8941	8910	8878	8847	8816	8785	8754	8724	5 10 16	21 26
49	·8693	8662	8632	8601	8571	8541	8511	8481	8451	8421	5 10 15	20 25
50	·8391	8361	8332	8302	8273	8243	8214	8185	8156	8127	5 10 15	20 24
51	·8098	8069	8040	8012	7983	7954	7926	7898	7869	7841	5 10 14	19 24
52	·7813	7785	7757	7729	7701	7673	7646	7618	7590	7563	5 9 14	18 23
53	·7536	7508	7481	7454	7427	7400	7373	7346	7319	7292	5 9 14	18 23
54	·7265	7239	7212	7186	7159	7133	7107	7080	7054	7028	4 9 13	18 22
55	·7002	6976	6950	6924	6899	6873	6847	6822	6796	6771	4 9 13	17 21
56	·6745	6720	6694	6669	6644	6619	6594	6569	6544	6519	4 8 13	17 21
57	·6494	6469	6445	6420	6395	6371	6346	6322	6297	6273	4 8 12	16 20
58	·6249	6224	6200	6176	6152	6128	6104	6080	6056	6032	4 8 12	16 20
59	·6009	5985	5961	5938	5914	5890	5867	5844	5820	5797	4 8 12	16 20
60	·5774	5750	5727	5704	5681	5658	5635	5612	5589	5566	4 8 12	15 19
61	·5543	5520	5498	5475	5452	5430	5407	5384	5362	5340	4 8 11	15 19
62	·5317	5295	5272	5250	5228	5206	5184	5161	5139	5117	4 7 11	15 18
63	·5095	5073	5051	5029	5008	4986	4964	4942	4921	4899	4 7 11	15 18
64	·4877	4856	4834	4813	4791	4770	4748	4727	4706	4684	4 7 11	14 18
65	·4663	4642	4621	4599	4578	4557	4536	4515	4494	4473	4 7 10	14 18
66	·4452	4431	4411	4390	4369	4348	4327	4307	4286	4265	3 7 10	14 17
67	·4245	4224	4204	4183	4163	4142	4122	4101	4081	4061	3 7 10	14 17
68	·4040	4020	4000	3979	3959	3939	3919	3899	3879	3859	3 7 10	13 17
69	·3839	3819	3799	3779	3759	3739	3719	3699	3679	3659	3 7 10	13 17
70	·3640	3620	3600	3581	3561	3541	3522	3502	3482	3463	3 6 10	13 17
71	·3443	3424	3404	3385	3365	3346	3327	3307	3288	3269	3 6 10	13 16
72	·3249	3230	3211	3191	3172	3153	3134	3115	3096	3076	3 6 10	13 16
73	·3057	3038	3019	3000	2981	2962	2943	2924	2905	2886	3 6 9	13 16
74	·2867	2849	2830	2811	2792	2773	2754	2736	2717	2698	3 6 9	13 16
75	·2679	2661	2642	2623	2605	2586	2568	2549	2530	2512	3 6 9	12 16
76	·2493	2475	2456	2438	2419	2401	2382	2364	2345	2327	3 6 9	12 15
77	·2309	2290	2272	2254	2235	2217	2199	2180	2162	2144	3 6 9	12 15
78	·2126	2107	2089	2071	2053	2035	2016	1998	1980	1962	3 6 9	12 15
79	·1944	1926	1908	1890	1871	1853	1835	1817	1799	1781	3 6 9	12 15
80	·1763	1745	1727	1709	1691	1673	1655	1638	1620	1602	3 6 9	12 15
81	·1584	1566	1548	1530	1512	1495	1477	1459	1441	1423	3 6 9	12 15
82	·1405	1388	1370	1352	1334	1317	1299	1281	1263	1246	3 6 9	12 15
83	·1228	1210	1192	1175	1157	1139	1122	1104	1086	1069	3 6 9	12 15
84	·1051	1033	1016	0998	0981	0963	0945	0928	0910	0892	3 6 9	12 15
85	·0875	0857	0840	0822	0805	0787	0769	0752	0734	0717	3 6 9	12 15
86	·0699	0682	0664	0647	0629	0612	0594	0577	0559	0542	3 6 9	12 15
87	·0524	0507	0489	0472	0454	0437	0419	0402	0384	0367	3 6 9	12 15
88	·0349	0332	0314	0297	0279	0262	0244	0227	0209	0192	3 6 9	12 15
89	·0175	0157	0140	0122	0105	0087	0070	0052	0035	0017	3 6 9	12 14

N.B. — Numbers in difference-columns to be subtracted, not added.

ANSWERS

5. $\sqrt{l^2 + b^2 + h^2}$, dir. cosines $l : b : h$.

6. 5, $\theta = \tan^{-1}\frac{4}{3}$. $(90°.)$

   I or $-$ I. $\qquad (180°.)$

   7. $\qquad\qquad (0°.)$

7. o.

12. 5, 8.66.

23. (a) 1936; (b) 35.405;

   (c) 983.5.

24. 45+ mi. per hr.

26. 40 mi. per hr.

27. 96.56.

29. Area $2x \cdot 10$, $[x =$ instantaneous length of side.

   Volume $3x^2 \cdot 10$, $[x =$ instantaneous length of side.

30. 1162 m. If $t = 0°$C.

33. See Introduction I, "Dimensions."

34. 75 cm. per sec.²

35. $-$ 4015.7 km. per hr.²

36. 120 cm. per sec.

39. 1, 3, 5, $\cdots (2n - 1)$.

40. 234 cm.

41. (a) 1152 cm.; (b) 270 cm.

42. (a) 1264 cm.; (b) 284 cm.

44. $\frac{3.6}{7}$ km. per hr.²

45. 8th sec.

46. 48 km.

47. 27.5 hrs.

48. $V_0 = 0$; $a = 2$.

49. 3 sec.

50. (a) 8.66 sec.; (b) 3.54 sec.

56. 1600 dynes.

60. 500 sec.

64. $196 \cdot 10^4$ dynes.

65. $623 \cdot 10^5$ dynes.

66. Last part, 35280.

69. First, 1470; second, 22050 cm.

70. 49 kg.

72. 122.5 m.; 24.5 m. per sec.

73. 5.87 sec.

74. 4427+.

75. 36.3.

76. 44.1 m.

77. 90.4 m.

81. 10.4 m. up; 9.2 m. down.

83. 0.5 sec., nearly.

85. 20.4 m.

86. 2.04 sec.; 4.08 sec.

87. $416\frac{2}{3}$.

94. 485 cm. per sec.; 0.5 sec.

96. 5.83.

98. $6°55'$.

99. 913.8 cm. per sec.

102. 264 ft. per min.

103. $66\frac{2}{3}$ ft. per sec.

104. $\sqrt{2} : 1$.

105. 8.54 mi. per hr.; $57°25\frac{1}{2}'$ E. of S.

106. 7.071 mi. per hr.

107. 36.56 km. per hr.

108. 51.96 m. per min.; 30 m. per min.; 0.

109. 26.4 ft. per min.

110. 17.39; 12.30; 4.658; $-$ 2.

111. 30.53+, 71° with "a," nearly.

112. 47.1°; 3.219.

113. 19.05; 0; $-$ 22.

114. 8.659 sec.

115. 7.14 sec.

120. 326.53 m.; 653.1 m.

121. (1) 2.49 sec.; (2) 498.4 ft.; (3) 215 ft.; (4) $21°50'$, nearly.

124. (b) $\dfrac{2\pi}{60}$ .1000 radians per min.

125. Angular velocity alike; linear as 1 : 2.

127. 4π radians per sec.

129. 523.6 mi. per hr.(when $r = 4000$ mi.).

130. 33 : 8.

131. 25 m.; $39\frac{3}{5}$ m.; 0.

132. 4.1 grams.

134. 2.5 ft. per sec.

135. 131+ lb.

136. 10.35 kg. wt.; 4.35 kg. wt.

137. $T = 2g\,\dfrac{Mm}{M+m}$; $a = \dfrac{M-m}{M+m}$.

139. Uniform motion; $T = gM$.

140. $\frac{1}{2}g$.

142. $130\frac{2}{3} \cdot 10^5$ dynes.

144. 5.625, 4.375.

145. $a = \dfrac{Mg}{M+m}$; $T = \dfrac{mMg}{M+m}$.

146. $\dfrac{M}{m} = \dfrac{2}{3}$.

150. $53 \cdot 10^4$ dynes.

151. 15; 3; 14.5; 13.9; 10.82; 7.93; 4.84.

152. 0.7265.

153. 12.2; 37.4.

154. 2.

155. 60°.

156. 120°.

157. 0°.

162. 60°.

163. 4 kg. wt.

164. 7921.4 dynes; 15843 dynes.

166. $P \cdot$ sec. 10° lb. wt.

167. 11.5 ; 27.7.

168. 20; 20; 21.22.

169. 45° incl. Algebraic sum $= 282.8$ gr. wt.

175. 911+ cm.

176. 20000 lb. wt.

180. 10 cm.

183. $\frac{1}{2}ap + \frac{2}{5}P$; $\frac{1}{2}ap + \frac{3}{5}P$.

186. 3600.

187. 50.9 [kg. cm.].

197. (a) $\dfrac{l}{2}$;

(b) $\dfrac{\frac{1}{2}k\,[x_2{}^2 + x_1 x_2 + x_1{}^2] + \dfrac{\rho_0}{2}[x_2 + x_1]}{\frac{1}{2}k\,[x_2 + x_1] + \rho_0}$;

(c) $\frac{2}{3}h$.

209. 2000 ergs.

210. $216 \cdot 10^6$.

211. $98 \cdot 10^6$; $294 \cdot 10^6$.

213. $2352 \cdot 10^7$.

214. 16 m.

215. 100 m.

216. 34640.

217. $\dfrac{20000}{\cos 10°}$.

218. $5 \cdot 10^8$.

219. $2 \cdot 10^5$.

220. $32 \cdot 10^5$ gr. cm.

221. $96 \cdot 10^5$ gr. cm.

224. $72 \cdot 10^3$ kg. m.

226. $2 \cdot 10^6$ ergs.

228. $49 \cdot 10^{11}$; $24.5 \cdot 10^{11}$.

230. $W = mal$.

231. $\frac{1}{4}$ as large. $\therefore$ Numeric 4 times as great.

234. $4 \cdot 10^{16}$ ergs.

235. $588 \cdot 10^{10}$ ergs; $6 \cdot 10^4$ kg. m.

236. $588 \cdot 10^6$ ergs.

237. $M_1 = \frac{1}{16}M_2$; kinetic energy will be acquired by the system.

239. $41552 \cdot 10^6$ ergs.

240. $6272 \cdot 10^6$.

241. $197392 \cdot 10^4$; $49348 \cdot 10^4$.

242. $[11267 \cdot 10^5$ total energy]; 6.47 cm.

243. $591 \cdot 10^{11}$ ergs approx.

244. 4000 ergs.

245. $1125 \cdot 10^7$.

246. $27 \cdot 10^3$; $51 \cdot 10^3$.

248. $64 \cdot 10^7$ dynes.

249. $48 \cdot 10^5$ ergs.

250. $24 \cdot 10^4$.

251. (a) $25 \cdot 10^7$ m.; (b) $625 \cdot 10^6$ m.; (c) 0; (d) $625 \cdot 10^6$ m.

252. $980 \cdot$ m. ergs.

260. $45 \cdot 10^9$ ergs.

261. $5 \cdot 10^8$ ergs; 5.1 kg. m.

262. $\sqrt{2gh}$.

271. 0.199.

272. $49 \cdot 10^8$ ergs.

281. 10053 kg. wt.
285. 1000 : 1.
286. 6⅔ kg. wt.
287. 1 : 24.
289. 160 kg.
302. 98 · 10⁶.
309. 48 · 10¹².
314. − 320; 1600.
318. $\dfrac{d^2 V}{dx^2} = 0$.
332. .02.
333. 142+.
334. 44.8 tons.
335. 16°42′.
336. 28.62.
337. (1) .2.
339. Equate *resultant* force to $(M + L)a$, and solve for a.
341. $g\,[\sin 60° - \mu \cos 60°]$.
342. 2656 · 10⁴.
349. .8 sec. approx. $l = 16$;
      1.14 sec. approx. $l = 32$.
350. 802+.

351. $\frac{9}{18}$.
352. Ratio 1.000046.
353. $T' = \dfrac{T}{\sqrt{2}}$.
354. (a) .875; 1.43; (b) 1.253.
355.
356. 1.718 sec.
358. $\frac{1}{3} Ml^2$; $\frac{1}{12} Ml^2$.
359. (a) $\frac{1}{3}\rho_0 l^3 + \frac{1}{4} k l^4$; ·
      (b) $\frac{1}{3}\rho_0 l^3 + \frac{1}{12} k l^4$.
362. $\dfrac{Mr^2}{4}$ one-fourth mass × square of radius.
363. $\dfrac{Mr^2}{2}$ one-half mass × square of radius.
368. 392 · 10³.
369. 245 · 10⁵, increased fourfold.
371. 4 · 10⁻⁴.
372. 625 · 10⁻⁶.
376. 17 · 10¹¹.
377. .26 cm.
384. 113.54.

LIQUIDS AND GASES

393. About 3 A.
395. 96.4 gr. wt.
396. 123 approx.
398. 93.5 meters.
402. 5 : 3.
417. 40560 kg.
418. 97200 lbs.
419. 12.
420. $\dfrac{1}{13.6}$ of its height.
423. 21.5 : 11.3 : 8.9 : 2.6.
      46.5 : 88.5 : 112 : 383.
      3.6 : 4.45 : 4.96 : 726.
424. $V_g : V_s = .535$.
428. 257; 1426; 159.
429. 10.5 : 19.3; 2.66 : 2.15.
430. 4.
431. 40.

432. 32.
433. 137.6 gr. wt.
434. 2.
435. 1.6.
436. 1 : 2.
437. .6.
438. .2.
439. 735; 1470.
441. .5.
443. $\frac{9}{7}$.
445. 4.37.
446. Inversely as the densities.
447. 11.
448. 876.
449. 3.
450. .79.
451. 1.2.
452. .9.

453. (1) 286 gr. wt.; 313.5 gr. wt.

455. 2.9 $\left[\text{Note that } \delta = \dfrac{\Sigma V s}{\Sigma V} \right]$.

456. 4.84; 5.09.

459. $\frac{6}{7}$.

461. 11.3 c.c.

462. 18 % approx. in Hg.

464. 975 cm.

HEAT

477. 113°; 53.6°; − 4°.

478. 100°; 22.2°; 0°; − 34.4°.

480. − 40°.

481. 160°.

486. 12.618 m.

488. The increase in length is equivalent to 13.6 added terms.

492. 189 × 10⁻⁷.

493. 1129°.

504. $\frac{1}{35738}$.

505. 40.197 c.c.

507. 13.11.

508. 13.35.

509. $\frac{255}{1826}$.

513. 176.25°.

532. 26226.

533. As 11 : 21.

534. 781052.

535. As 3 : 55 nearly.

536. 27.6°.

537. 12.7 and 42.3 liters.

540. 5.78 grams.

541. 5.6 grams.

542. 4.91 grams.

543. .06.

544. .62.

545. 3.29 cm.

613. 4.9 grams.

614. 81363 cm. per sec.

618. 30618.75 calories.

619. 21851.7 calories.

620. 4.189 × 10⁷ ergs.

624. $W = A p' a \left[1 + \log_e \dfrac{l}{a} \right] - A p_e l.$

625. (a) 4386.3 ft.-lbs. (b) 47.85 H.P.

626. 4.2 H.P.

ELECTRICITY AND MAGNETISM

632. $F = .01$ dynes repulsive.

633. $F = .64$ dynes attractive.

634. $f' = 4 f$; $r' = 2 r$.

635. $q = 25.6$.

636. $r' = 2 r$.

640. Surface density $= \dfrac{1}{4 \pi}$.

643. 8000 dynes.

654. (a) $V' = 4 V$; $V' = − V$.

657. − 42.

662. $Q = 10000$; V = 1000; force =
100.

663. Work = 1800 ergs.

664. 80 ergs.

670. V = 2; $f = 0$.

688. $f = 12.5$.

689. V = 10000.

693. 4 cm.

695. (a) 50000; (b) 5000; (c) 500.

697. 1600 ergs.

703. Loss $\frac{3}{4}$.

707. $K'' = 6.5$.

711. Energy = 1 : 6.

712. V and Q reduced $\frac{1}{4}$ initial values.

719. $\frac{3}{8}$ energy remains.

720. $\dfrac{100000}{\pi}$; $C = 31847$.

721. 15.9 · 10⁷ ergs.

725. $\frac{1}{2}$ W. each jar.

729. Cap. = 7; change in large sphere
= 21.43; small sphere 8.57; energy over wire 18.57 units; initial
energy = 185 ; final energy =
64.3; final potential = 4.29.

731. $I = 2$ amperes; 600 coulombs.

732. 8 amperes.

733. $R' = 3R$.

734. 5 amperes.

735. 4800 coulombs.

742. (a) 450; (b) 900.

746. (a) .0377 amperes; (b) .377, .754, 1.131, 1.808; (c) 3.77 volts.

755. 8 ohms.

758. $R_i = 25$ ohms.

759. 2.58 volts.

762. $1\frac{4}{13}$ volts.

767. Radius doubled.

768. 10000 ohms.

770. 1.66 ohms.

771. (a) 1.19; (b) 14; (c) 7.79; (d) 140.

772. $R_2 = 35.26$ ohms.

773. 3; $1\frac{3}{7}$; $4\frac{1}{2}$ ohms.

774. Length $= \frac{1}{8} l$.

775. $253\frac{1}{3}$ ohms.

776. $59\frac{18}{65}$ ohms.

777. 27 ohms.

778. $x = 1111.11$ ohms.

781. Take intersection of line $\dfrac{x}{r_1} + \dfrac{y}{r_2}$
$= 1$ with $x = y$.

782. $E = 12$; $r_1 = 72$ series; $E = 2$; $r_1 = 2$ multiple.

783. $E = 12$ volts; $R = 10\frac{1}{2}$ ohms.

784. .8 ohms.

789. 6; 3; 4; 13.

790. 3 : 1.

791. (a) 30 volts; (b) 59 ohms; (c) .508+ amperes; (d) 16.3 volts; (e) 5.54 (A to B).

792. (a) 8.332 ohms; (b) 12 amperes; (c) 100 volts; (e) 111.96 volts.

795. .028 amperes in branch 10.

798. 1 %.

820. 6×10^5 joules.

824. 1008 ohms; 10080 volts; 600 coulombs passing per min.

826. $28.8 \cdot 10^5$ joules; $6.9 \cdot 10^5$ calories.

829. $I = 10.04$ amperes; 10040 volts.

837. $IV = 64000$ ergs; $I^2R = 16000$ ergs.

840. 45.11 ohms.

841. .126 L.

846. $R_{40} = 1021.2$; $R_{80} = 1042.4$.

847. 256° C.

848. 2187.2 ohms.

850. 7.7 ohms at 0°.

853. 2.362 g. of copper.

855. 5 amperes.

864. 1.9017 amperes; .026 amperes; $14.36 \cdot 10^{-6}$ amperes.

868. (a) radius $= 157$ cm.; (b) $\delta = 470$.

869. Total current $= .0838$ amperes.

872. $I_0' = 2.81 \cdot 10^{-3}$.

873. .1 amperes.

874. $I_0' = 13.3 \cdot 10^{-6}$.

881. $I_0 = 138$.

882. $I_0 = .0225$.

883. $I_0 = 38.6 \cdot 10^5$.

884. $l = 7p$.

885. 138 cm.

886. 490 amperes.

887. Force $= 4\sqrt{2}\,\dfrac{k}{a}$.

899. Force ∥ to bar equals .29 dynes.

900. 3912 dynes.

901. $M = 4.66$ C.G.S. units.

904. $H = 8$.

907. $H = .208$; $V = .534$.

911. $M = 546$.

913. $M = 6000$.

914. (a) 102; (b) 72.114.

916. 140.

918. 1.2.

919. 120 ergs.

920. $V = 8.33$.

944. $k = 30.65$; $B = 15438$; $\mu = 386$.

947. $B = 3508$.

949. 2524.

950. 134.3.

954. (a) $H = 125.7$ per sq. cm.

956. 751.1 watts.

960. $M = 214.765$.

962. .141.

983. 6 dynes.

987. 1.8 volts.

988. 1000 dynes; 100000 dynes.

989. $5 \cdot 10^5$ dynes ⊥ to field.

990. $13.76 \cdot 10^7$ coulombs.

R

994. (*a*) $8.5 \cdot 10^{-4}$ volts; (*b*) $8.38 \cdot 10^{-4}$
    volts; (*c*) 297.7 sec.
1000. .32 volts.
1031. 15.77 H.P.
1032. Electrical eff. 92.6 %.
1033. Electrical eff. 83.3 %.
1034. 62.5 amperes.
1035. 96.5 %.
1036. 4.5 H.P.
1041. $12 \cdot 10^6$ dynes.
1046. Net eff. 91.2 %.

1047. 77 %.
1051. 76.44 amperes.
1052. 100 amperes.
1054. Max. $I = 49.93$ amperes; mean
    value = 31.3 amperes.
1059. 843; 904.3.
1061. Imp. = 454.
1062. $\theta = 87°34'$.
1063. 3300 watts.
1066. $L = .024$ henrys.
1069. 100 volts; 36 amperes.

SOUND AND LIGHT

1094. .419; 6; 77.4.
1095. 8; 160; − 1600; .314.
1101. 2 crests, 3 troughs.
1102. 20.
1114. (1) .6283; (2) 1.257; (3) 2.
1115. $y = a \sin \pi [8t + x]$.
1124. 332 m.
1128. (1) 27.7 m.; (2) 55.4; (3) 83.
1129. $\dfrac{x}{v}$.
1130. 34740 cm.
1132. 3444 m.
1133. $\lambda_{20} = \lambda_0$ 1.036.
1134. 500 waves.
1135. 23.7° C.
1139. 135.2 cm.
1140. Velocity and wave length in-
    creased.
1141. 328 m.
1147. 128; 362.1; 181.
1148. (1) Make string $\frac{4}{5}$ of its former
    length; (2) increase tension
    by the factor 1.26+.
1150. $42 \cdot 10^6$ dynes.
1156. $\dfrac{F_1}{F_2} = \left[\dfrac{4}{9}\dfrac{.0098}{.0045}\right] = .968$.
  or $\dfrac{F_1}{F_2} = \left[\dfrac{1}{9}\dfrac{.0098}{.0045}\right] = .242$.
1157. $\dfrac{l_1}{l_2} = 1.478$.
1158. $\dfrac{F_1}{F_2} = \left[\frac{9}{16}\text{ of } 2.1778\right] = 1.225$.

1172. 240 cm.
1173. 80 cm.
1174. 26.6 cm.
1175. 145; 435; 1305.
1177. 120 cm.; 170.
1179. At 0° length 50 cm.; at 25°
    length 52.2 cm.
1180. 192; 320; 448; 576.
1181. 128; 192; 256; 320.
1182. 2.1 cm.; 6.2 cm. from wall.
1183. 18.7 cm.
1184. 5 beats.
1189. $n = 208$; $n = 1040$.
1190. 2.8; 8.5; 70.8 cm.
1192. 8.3 cm.; 7.9 cm.; 31.9 cm.
1215. 10 cm.
1216. 13.3 cm.; 14; 16.7 etc.
1217. 1.7 cm.; 2.0 cm.; 3.3 cm. etc.
1220. $\frac{3}{4} R$; $\frac{1}{4} R$.
1221. $R = 15$ ft.
1225. $\dfrac{R}{2}$; from natural size to zero.
1227. 15.6 cm. per. sec. toward.
1235. 41°49'.
1236. 1.3214; .7567.
1237. 74°37'.
1238. Angle of refraction = 11°12'.
1239. 32°2'; 40°30'; 46°25'.
1240. 70°32'.
1243. 20°11'.
1244. Yes; critical angle increases with
    increase of wave length.

1245. $225 \cdot 10^8$; $200 \cdot 10^8$; $185 \cdot 10^8$.

1246. 165 sec.

1247. Angle of refraction in glass
      $= 25°40'$.

1248. 1.07

1249. 1.33 m.

1250. .582 cm.

1251. 40 ft.

1252. 40 ft.

1253. Above.

1258. $\mu = 1.668$.

1259. $23°38'$; $10°22'$.

1260. For yellow light taking index of
      crown glass as 1.530,
      $\delta_w = 10°22'$; $\delta_g = 16°40'$.

1270. Taking 1.530 as index, 1.24 in.
      and 7.44 in.

1274. 12 m.; 7.5 m.; 4.8 m. etc.

1280. 30 cm.

1281. 100 cm.

1282. .9 cm.

1285. $\mu = 1.5$.

1286. $f =$ radius.

1302. $312 \cdot 10^{-7}$ cm.

1303. $76 \cdot 10^{-6}$ sec.

1309. $22 \cdot 10^{-8}$; $38 \cdot 10^{-8}$ ... etc.

1310. 1.21 : 1.

1318. $3°23'$; $6°47'$; $10°12'$.

1319. 2 : 3.

1322. 1059.

INDEX

ACCELERATION, 38, 39, 40, 41.
Approximations, 33.
Archimedes' principle, 94.
Atmospheric pressure, 89.
Averages, 31.

BAROMETER, 89.
Batteries, 137, 138, 139, 142, 143.
  best arrangement of, 146, 147.
Boiling points, table of, 16.
Boyle's law, 98.

CALORIE, 108.
Calorimeters, 109, 110, 111, 112, 113.
Capacity, electrical, 124.
  specific inductive, 18, 128.
  thermal, 108, 109.
Cells, best arrangement of, 147.
  grouping of, 142, 143.
Center of inertia, 58, 61.
  of mass, 58, 61.
  of gravity, 58, 61.
Coefficient, of expansion, 101.
  cubical, 103, 104, 105.
  of gases, 106, 107, 108.
Coefficients of expansion, 16.
Condensers, 128, 129.
Conductivities, thermal, 17.
Critical angle, 213.
Current alternating, 183, 184, 186.
Current electricity, 132.

DENSITIES, tables of, 14.
Diffraction grating, 222, 223.
Dimensions, 5, 187–190.
Dimensional equations, 5.
Doppler's principle, 207.

Dynamo, 179–183.
  characteristic curve, 180, 181.
  efficiency, 181, 182.

ENERGY, of charge, 127.
  of discharge, 130, 131.
  kinetic, 67.
  of rotation, 67, 74.
  transformation to potential, 75–78.
Elastic limit, 85.
Elasticity, 85.
Electric force, 123.
Electrochemical equivalent, 153, 154.
Electromagnetic units, 18, 187–190.
  attraction, repulsion, 169–172.
  induction, 169–182.
Electromotive force, 142.
  of induction, 173–178.
Expansion coefficients, 16.

FALL, of potential in a wire, 135.
  of potential and electromotive force,
    135.
Farad, 18.
Faraday's disc, 177.
Fields of force, electric, 124, 125.
  magnetic, 161, 169, 173.
Force, 40–43.
  systems, 53–57.
Friction, 79.
  angle of, 79, 80.
  coefficient of, 80.

GALVANOMETER, 155, 156.
  Ballistic, 157.
Gases, 89 et seq.
Graphic methods, 27, 30.

HEAT, 100.
  specific, 108.
  specific variation of, 110.
  of fusion, 110–112.
  of vaporization, 112, 113.
  in electric circuit, 149–151, 183.
Heats, of liquefaction, table of, 15.
  of vaporization, table of, 16.
Hydrometers, 97.
Hydrostatic pressure, 90–92.
  press, 92, 93.
Hysteresis, 167.

INDICATOR diagram, 119.
Indices of Refraction, 20.

KILOGRAM, 8, 9.
Kirchhoff's law, 147–149.

LENSES, 216.
  images by, 217, 218.
  curvature of, 217–219.
Light, reflection of, 209.
  velocity of, 213, 214.
  refraction of, 212–215.
  interference of, 220–223.
  diffraction of, 222, 223.
Lines of force, 122.
  magnetic, action of, 168–170.
Liquids, and gases, 89.
  pressure, 89–91.

MAGNETIC, field, due to currents, 159.
  induction, 165.
Magnetism, 161.
Magnetization curve, 165, 166.
Magnetometer, 163, 164.
Mass and weight, 7.
Measurement, 1.
Mechanical equivalent of heat, 14.
Melting points, table of, 15.
Mirrors, 209.
  plane, 209, 210.
  concave, 210, 211.
  convex, 211, 212.
Moment of inertia, 82–84.
Motor, 183.

Multiple resistance, 139–141.
  graphic methods, 141.

NEWTON's rings, 221.

OHMS, various, 18.
Ohm's law, 132 et seq.
Overtones, 205, 206.

PENDULUM, gravity, 82.
  magnetic, 167, 168.
Potential, diagrams, 135–138.
  gravitational, 75–78.
Prism, 214, 215.
Pressure, of atmosphere, 89.
  of gases, 98, 99, 106–108.
  of liquids, 89, 92.
  center of, of liquids, 92.
Projectiles, 49.
Pulleys, systems of, 73, 74.

REFRACTION, index of, 212–215.
  indices of, 20.
  law of, 212.
Resistance, multiple, series, 138, 149.
  specific table of, 17.
  temperature coefficients of, 17.
  units of, 18.

SELF-INDUCTION, 185.
Shunts, 143–145.
Simple harmonic motion, 191–193.
Sound, 198.
  musical, 200.
  velocity of, 19, 199.
Specific gravity, 94–97.
  gravity bottle, 96.
  heats, 14, 15.
  inductive capacity, table of, 18.
  resistance, 152.
Spectra, 215, 222, 223.
Static electricity, 121.
Strain, 85, 86.
Stress, 85, 86.

TEMPERATURE, 100.
Thermometer, 100.
  scales, 101.

Thermometer weight, 105, 106.
Thin plates, 220, 221.
Torsion, 87, 88.
    moment of, 88.
Transformer, 186.
Transmission of energy, electric, 151.

Units, 1.
    C. G. S. and practical, 190.
    electrical, magnetic, 187.
    fundamental and derived, 2.
    of area, 12.
    of force, 13.
    of heat, 16.
    of length, 12.
    of mass, 13.
    of power, 13.
    of resistance, 18.
    of stress, 13.
    of volume, 12.
    of work, 13.
    practical, in C.G.S., 18.
    transformation of, 190.

Vapor, pressure, 114, 118, 119.
    volume, 114, 118, 119.

Vectors, 21.
    addition of, 21, 22.
    examples on, 23, 25.
Velocity, of light, 19.
    of sound, 19, 199.
    of sound, temperature, effect on, 199.
Vibration, 191.
    columns of air, 205–207.
    elliptic, 192.
    strings, 201–204.

Wave length of sound, 200.
    of light, 221, 223.
Wave lengths of light, table of, 19.
Waves, 194, 195.
    phase, 197.
    progressive, 196.
    retardation of, 197.
    sound, 198.
Wheel and axle, 74, 75.
Work, by torque, 65.
    constant force, 62.
    general expression for, 69, 70.
    principle of, applied to machines, 71.
    variable force, 63, 69.

Young's modulus, 86.